Chemical Engineering

Solved Problems

N.S. Nandagopal, PE

Professional Publications, Inc. • Belmont, CA

How to Locate and Report Errata for This Book

At Professional Publications, we do our best to bring you error-free books. But when errors do occur, we want to make sure you can view corrections and report any potential errors you find, so the errors cause as little confusion as possible.

A current list of known errata and other updates for this book is available on the PPI website at **www.ppi2pass.com/errata**. We update the errata page as often as necessary, so check in regularly. You will also find instructions for submitting suspected errata. We are grateful to every reader who takes the time to help us improve the quality of our books by pointing out an error.

CHEMICAL ENGINEERING SOLVED PROBLEMS

Current printing of this edition: 1

Printing History

edition number	printing number	update
1	1	New book.

Printed in the United States of America

PPI
1250 Fifth Avenue, Belmont, CA 94002
(650) 593-9119
www.ppi2pass.com

Library of Congress Cataloging-in-Publication Data
Nandagopal, N. S., 1955-
Chemical engineering solved problems / N.S. Nandagopal.
p. cm.
ISBN-13: 978-1-59126-090-5
ISBN-10: 1-59126-090-6
1. Chemical engineering--Problems, exercises, etc. 2. Chemical engineering--Examinations, questions, etc. I. Title.

TP168.N367 2007
660.076--dc22

2006051003

Table of Contents

Preface

The goal of *Chemical Engineering Solved Problems* is to enhance your preparation and readiness for the Principles and Practice of Engineering (PE) examination in chemical engineering. The problems included in *Chemical Engineering Solved Problems* are original; I developed them during several years of teaching review classes for the chemical engineering PE exams. The originality of the problems respects the confidentiality of the NCEES exams, which is of great importance to me and to PPI.

NCEES describes the subtopics in each exam subject area as "knowledge clusters." For example, within the subject area of heat transfer, the knowledge clusters are heat exchanger design and performance; energy conservation; conduction, especially insulation problems; convection; radiation, especially furnace design; and evaporation. The problems included in *Chemical Engineering Solved Problems* cover the breadth of knowledge clusters specified by NCEES. This was an important and challenging goal for me while developing the problems. This book will expose you to a variety of questions, both conceptual and practical, covering the broad range of topics you will encounter in the actual exam. By working through the problems presented here, you will reinforce your fundamentals in the knowledge clusters encountered in the NCEES PE exam.

Detailed solutions are presented for all the problems included in *Chemical Engineering Solved Problems*. While developing these solutions, I have used standard sources for physical properties and other data.

Although great care was exercised in preparing this book, occasional errors may still be present. I would appreciate your feedback in pointing out the errors. Please feel free to provide your comments and suggestions for improving this book through PPI's online errata submission form at **www.ppi2pass.com/errata**.

To ensure maximum benefit from this book, I urge you to first read the Introduction.

Good luck in the examination!

N. S. Nandagopal, PE
Houston, Texas

Acknowledgments

I would like to express my appreciation to the staff members at Professional Publications, Inc. who were involved in producing *Chemical Engineering Solved Problems*. In particular, I would like to thank Aline Magee, who guided me with great dedication and professionalism through different stages of the book. I would also like to thank Michael R. Lindeburg, President of PPI, and Tom Tolfa for providing me the opportunity to write this book.

Dr. Haku Israni of Test Masters Educational Services, Inc., Houston, Texas, has provided me with many opportunities to teach FE and PE review classes. I developed the expertise necessary to write this book solely because of the experience I gained while teaching these review classes. Therefore, my thanks go out to Dr. Israni.

I also want to recognize the patience and support of my wife, Lakshmi, and my son, Nikhil, while I worked on this book. In particular, my wife has expended great efforts in producing the manuscript. For this, I am very grateful.

Finally, my deepest gratitude to the good Lord for providing me with the ability, opportunity, and fortitude necessary to accomplish this task.

N. S. Nandagopal, PE
Houston, Texas

Introduction

ORGANIZATION OF THIS BOOK

Although NCEES does not release old PE exams or reveal specific exam problems, it has identified the exam subject areas and their approximate weighting as a percentage of the exam questions. *Chemical Engineering Solved Problems* is organized into seven sections that correspond to the sections of the Chemical Engineering PE exam.

Subject Area and Knowledge Clusters	Approximate Percentage of Examination

I. MASS/ENERGY BALANCES AND THERMODYNAMICS **24%**

A. *Mass Balances* (11%):
material balances and stoichiometry; phase behavior; process variants: bypass, recycle, and purge; combustion processes (e.g., water-free analysis, excess air, staged combustion)

B. *Energy Balances and Thermodynamics* (13%):
sensible heat (heat capacity); latent heat (e.g., fusion, vaporization, sublimation); heat of reaction (exothermic, endothermic); heat of solution; estimation and correlation of physical properties; applications requiring combinations of sensible heat calculations, latent heat considerations, heats of reaction, etc.

II. FLUIDS **17%**

A. *Fluid Transport* (3%):
physical properties (e.g., viscosity, density, surface tension); pipe and tubing data (e.g., schedule number, surface roughness)

B. *Mechanical-Energy Balance* (11%):
potential (e.g., elevation change) and kinetic energy (e.g., velocity); friction: Reynolds number, pressure drop (e.g., friction factor, pipes, valves, fittings, expansion and contraction); flow applications: single conduit, parallel and branched systems, pumps, turbines, and compressors (e.g., work/energy requirements, efficiency and performance curves), two-phase flow (e.g., slug flow), filtration

C. *Flow Measurement Techniques* (3%):
pitot tube, orifice, and venturi, etc.; pressure differential measurement (e.g., manometers); mass flow (e.g., Coreolis, vortex shedding, thermal); permanent pressure drop (e.g., orifice valve)

III. HEAT TRANSFER **16%**

A. *Mechanisms* (6%):
physical properties (viscosity, density, heat capacity, etc.); conduction (e.g., Fourier's law in differential and integral form, parallel and series arrangements, mean area); convection: free (natural) convective heat-transfer coefficient, forced convective heat-transfer coefficient (metallic and nonmetallic); phase change (e.g., vaporization, condensation, sublimation, crystallization); combinations of mechanisms: (conduction, convection, and radiation in series)

B. *Applications* (10%):
insulation (e.g., type, sizing, and placement); measurement instruments (thermocouples, thermometers, RTD, IR, etc.); heat exchangers: overall heat-transfer coefficient, fouling factors, Reynolds number, mean temperature difference (LMTD, f-factor), types (e.g., double pipe, shell-and-tube, extended surface, plate), design (e.g., area, configuration, pressure drop), evaluation of existing and new exchanger systems (NTU method/pinch technology); service use of heat transfer equipment (e.g., condensers, reboilers, heat pumps); radiant and convective transfer

IV. MASS TRANSFER **13%**

A. *Phase Equilibria* (5%):
equilibrium data (e.g., VLE, LLE): equations of state, Henry's law and Raoult's law, non-ideal solutions (e.g., activity coefficient), azeotrope systems; phase equilibrium calculations: bubble and dew points, flash calculation; diffusion (e.g., purification, water treatment, chip manufacturing, chemical vapor deposition)

B. *Mass Transfer Contactors (Absorption, Stripping, Distillation, Extraction)* (7%):
continuous contacting (packed): minimum rate of flow of liquid (absorption), vapor (stripping), solvent (extraction), and reflux (distillation),

minimum number of transfer units or stages, height and number of transfer units or stages, types of packing, flooding—calculation of minimum vessel diameter, feed location for distillation column/tower; trayed contactors: minimum rate of flow of liquid (absorption), vapor (stripping), solvent (extraction), and reflux (distillation), minimum number of stages, theoretical stages—graphical methods, flooding—calculation of minimum vessel diameter, stage efficiency, feed location for distillation column/tower

C. *Miscellaneous Separation Processes* (1%): drying; adsorption (e.g., PSA, water treatment)

V. KINETICS 11%

A. *Reaction Parameters* (2%): rate constant; chemical equilibria; activation energy

B. *Reaction Rate* (2%): rate equation; order of reaction; analysis of experimental data from reaction systems

C. *Reactor Design and Evaluation* (5%): batch reactor; continuous stirred-tank reactor to include recycle to the reactor; plug-flow reactor (e.g., gas phase reactor); multiple reactors in series; yield and selectivity

D. *Heterogeneous Reaction Systems* (2%): multi-phase reactors: fluidized beds, packed beds; stability/runaway reactions; mixing

VI. PLANT DESIGN AND OPERATION 19%

A. *Economic Consideration* (2%): equipment-cost correlations (e.g., cost indices)/ economic calculations; operating costs; time value of money

B. *Design and Operation* (6%): process equipment design; process flow sheet development; design optimization; operating manuals (e.g., startup, shutdown, maintenance); equipment testing, troubleshooting, and analysis

C. *Safety* (5%): emergency venting devices (e.g., safety valves, blowout walls); performance of scheduled audits (e.g., testing safety valves, checking rupture, disks); flares and vents; plant layout considerations (e.g., equipment arrangement, pipe racks, and layouts); fire protection; emergency ingress and egress; process hazard analysis

D. *Environmental* (2%): evaluation and permitting of gas discharges and liquid discharges; solid waste management (non-hazardous and hazardous); industrial hygiene (e.g., MSDS, TLV, noise control, ventilation, personal protective equipment); pollution prevention

E. *Materials* (2%): materials properties and selection; structural design considerations (e.g., temperature limits, pressure limits, thermal expansion, pressure vessels per ASME Section VIII); corrosion considerations

F. *Process Control* (2%): sensors (e.g., choice, location); controller actions; feed-back/feed-forward actions; data interpretation

TOTAL 100%

HOW TO USE THIS BOOK

Chemical Engineering Solved Problems should be used to practice solving problems in each of the subject areas covered on the PE exam.

To optimize your study time and obtain the maximum benefit from the practice problems, use the following four-step process.

step 1: Review the problems in each subject area and identify those with which you are least familiar. Work a few of these problems to assess your general understanding of the subject and to identify strengths and weaknesses. When working problems, always make your best attempt to solve the problem before looking at the solutions provided in the book. Then use the solutions to check your work for those problems you are able to solve or to provide guidance in finding solutions to the more difficult problems.

step 2: Focus first on solving problems in those topic areas where the least mastery and greatest weaknesses exist. Begin by locating relevant resource materials, and then work the problems in one subject area at a time. As you work problems, some of these resources will emerge as being more important to you than others. These are the ones you will want to prepare for use when taking the PE exam.

step 3: Use the solved problems as a guide to understanding the general approach to solving problems in each subject area. Although the problems encountered on the PE exam may not be exactly the same as those presented in this book, the approach to solving problems will be the same. It may be useful to make notes in the margins to explain concepts or to reference sources that can help you solve a particular problem. These notes can be consulted while you are taking the exam.

step 4: After you have identified and addressed weaknesses by working problems in those subject areas with which you are least familiar, follow the same procedures outlined above, organizing resource materials and making notations, to work the problems in the remaining subject areas.

Remember that the solution presented for each example problem may represent only one of several methods for obtaining a correct answer. It may also be possible that an alternative method of solving a problem will produce a different, but nonetheless appropriate, answer.

EXAM ORGANIZATION

The PE exam consists of two parts: a morning session and an afternoon session, each lasting four hours. In completing the exam you will need to provide answers to 80 multiple-choice problems, 40 in the morning and 40 in the afternoon. Four possible answers are provided for each problem, but only one of the options is correct. The problems presented in *Chemical Engineering Solved Problems* follow a similar format. On the PE exam, one point is awarded for each correct answer, but no penalty is assessed for incorrect answers. Therefore, after you have answered problems you are able to solve, you should guess at those you are unable to solve. Don't leave any questions unanswered.

Both the morning and afternoon sessions of the exam are open book. In general, any bound reference material is allowed, including personal notes and sample calculations. Textbooks, handbooks, and other professional reference books are allowed. However, no writing tablets, scratch paper, or other unbound notes or materials are permitted. Battery-operated, silent, nonprinting calculators are allowed. However, each local jurisdiction defines specifically what materials you are permitted to bring with you to the exam. To find out what materials are allowed in the exam in your area, call your state board of registration.

The exam is meant to assess your personal competence without consultation, discussion, or sharing of information with others during the exam period. Do not expect to share any references or to communicate with others while taking the exam.

Information about the state boards can be found at **www.ppi2pass.com/stateboards.html**. More information about the chemical engineering PE exam can be found at **www.ppi2pass.com/chfaqs.html**.

Topics

Mass and Energy Balances

Thermodynamics

Fluids

Heat Transfer

Mass Transfer

Kinetics

Plant Design

Mass and Energy Balances

PROBLEM 1

A dilute soap solution containing 12% detergent is to be converted into a concentrated solution containing 42% detergent. A single evaporation process is used. However, the available evaporator achieves a concentration of 58% detergent in the stream leaving the evaporator. The desired product concentration is achieved by a portion of the dilute feed bypassing the evaporator. The feed rate of the dilute solution is 1000 lbm/hr (before bypass).

1.1. What is the production rate of the concentrated detergent solution?

(A) 120 lbm/hr
(B) 190 lbm/hr
(C) 290 lbm/hr
(D) 350 lbm/hr

1.2. What is the rate of evaporation of water?

(A) 620 lbm/hr
(B) 710 lbm/hr
(C) 810 lbm/hr
(D) 910 lbm/hr

1.3. What fraction of the feed bypasses the evaporator?

(A) 0.10
(B) 0.20
(C) 0.30
(D) 0.40

1.4. What is the feed rate to the evaporator?

(A) 600 lbm/hr
(B) 700 lbm/hr
(C) 800 lbm/hr
(D) 900 lbm/hr

1.5. Without the bypass stream, what would be the rate of evaporation of water?

(A) 700 lbm/hr
(B) 800 lbm/hr
(C) 900 lbm/hr
(D) 1000 lbm/hr

PROBLEM 2

Formaldehyde is produced by oxidizing methane using air. A side reaction is the combustion of methane to form carbon dioxide. The reactions are as follows.

$$CH_4 + O_2 \longrightarrow HCHO(g) + H_2O(g)$$

$$CH_4(g) + 2O_2 \longrightarrow CO_2 + 2H_2O(g)$$

100 mol/h of CH_4 at 25°C is fed to a reactor, and there is a 30% (mol %) conversion to formaldehyde. 20% of the remaining CH_4 participates in the side reaction to form carbon dioxide. Air at 100°C is fed at the rate of 500 mol/h. The products from the reactor are at 150°C.

2.1. What is the molar flow rate of nitrogen in the feed stream?

(A) 100 mol/h
(B) 200 mol/h
(C) 300 mol/h
(D) 400 mol/h

2.2. What is the molar flow rate of methane in the product stream?

(A) 44 mol/h
(B) 56 mol/h
(C) 62 mol/h
(D) 70 mol/h

2.3. What is the molar flow rate of water vapor in the product stream?

(A) 35 mol/h
(B) 41 mol/h
(C) 49 mol/h
(D) 58 mol/h

For Probs. 2.4 through 2.7, use elements at 25°C as the reference state for enthalpies. Assume that the products leave the reactor at 150°C.

2.4. What is the molar enthalpy of oxygen in the feed stream?

(A) 1.7 kJ/mol
(B) 1.9 kJ/mol
(C) 2.2 kJ/mol
(D) 2.5 kJ/mol

2.5. What is the molar enthalpy of formaldehyde in the product stream?

(A) −121 kJ/mol
(B) −119 kJ/mol
(C) −115 kJ/mol
(D) −111 kJ/mol

2.6. What is the molar enthalpy of water vapor in the product stream?

(A) −320 kJ/mol
(B) −290 kJ/mol
(C) −250 kJ/mol
(D) −240 kJ/mol

2.7. What is the rate at which heat must be removed from the reactor?

(A) 15 100 Btu/h
(B) 16 300 Btu/h
(C) 17 500 Btu/h
(D) 18 100 Btu/h

PROBLEM 3

The combustion product from a hydrocarbon fuel has the following Orsat analysis.

$$CO_2 = 13\%,\ CO = 0.5\%,\ O_2 = 3.5\%,\ N_2 = 83\%$$

3.1. How many moles of water vapor are formed per mol of fuel burned?

(A) 9.82 mol
(B) 10.6 mol
(C) 11.1 mol
(D) 12.2 mol

3.2. What is the molecular formula of the fuel?

(A) $C_{10.6}H_{20.2}$
(B) $C_{11.2}H_{22.4}$
(C) $C_{13.5}H_{21.3}$
(D) $C_{14.2}H_{26.4}$

3.3. What is the air-fuel ratio on a mass basis?

(A) 10 kg air/kg fuel
(B) 12 kg air/kg fuel
(C) 15 kg air/kg fuel
(D) 17 kg air/kg fuel

3.4. What is the percent excess air supplied?

(A) 17%
(B) 19%
(C) 22%
(D) 26%

3.5. If the total pressure is 1 atm, what is the dew point of the combustion product?

(A) 112°F
(B) 122°F
(C) 131°F
(D) 140°F

PROBLEM 4

Methane reacts with oxygen to form carbon dioxide and water. 200 lbmol/hr of a feed consisting of 25% methane, 65% oxygen, and 10% carbon dioxide are fed to a reactor that achieves a 90% conversion of the limiting reactant.

4.1. What is the percent excess of the excess reactant?

(A) 15%
(B) 23%
(C) 30%
(D) 36%

4.2. What is the molar flow rate of CH_4 in the product stream?

(A) 3 lbmol/hr
(B) 5 lbmol/hr
(C) 7 lbmol/hr
(D) 9 lbmol/hr

4.3. What is the molar flow rate of the product stream?

(A) 100 lbmol/hr
(B) 200 lbmol/hr
(C) 300 lbmol/hr
(D) 400 lbmol/hr

4.4. What is the mole fraction of CO_2 in the product stream?

(A) 0.20
(B) 0.25
(C) 0.28
(D) 0.33

4.5. If the total pressure is 14.7 psia, what is the partial pressure of oxygen?

(A) 2.9 psia
(B) 3.6 psia
(C) 4.2 psia
(D) 4.9 psia

4.6. If the temperature of the product gas is 360°F, what is the volume flow rate of the product gas?

(A) 1600 ft^3/min
(B) 1800 ft^3/min
(C) 2000 ft^3/min
(D) 2300 ft^3/min

Mass and Energy Balances Solutions

SOLUTION 1

1.1. A schematic for the process is shown, and the following nomenclature is used.

D: detergent
W: water
F: feed
B: bypass stream
P: product stream
E_o: evaporator outlet stream
E_i: evaporator inlet stream
H: water evaporated

The mass balance for the detergent is

$$\text{input} = \text{output}$$

$$\left(1000\ \frac{\text{lbm}}{\text{hr}}\right)\left(0.12\ \frac{\text{lbm } D}{\text{lbm}}\right) = P\left(0.42\ \frac{\text{lbm } D}{\text{lbm}}\right)$$

$$P = \boxed{285.7\ \text{lbm/hr}}$$

The answer is C.

1.2. The overall mass balance is

$$\begin{aligned} F &= P + H \\ H &= F - P \\ &= 1000\ \frac{\text{lbm}}{\text{hr}} - 285.7\ \frac{\text{lbm}}{\text{hr}} \\ &= \boxed{714.3\ \text{lbm/hr}} \end{aligned}$$

The answer is B.

1.3. Junction C is isolated as shown.

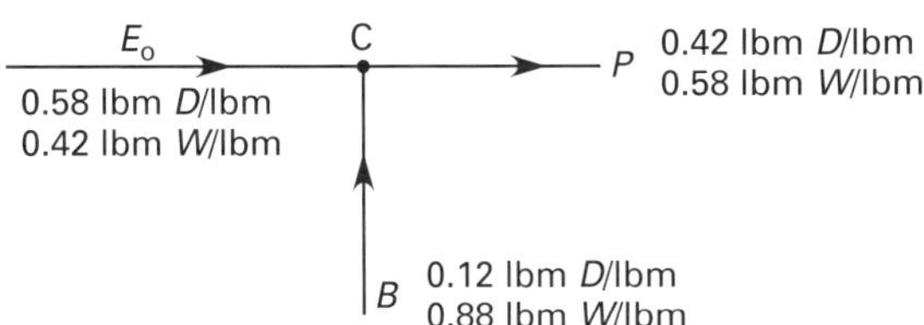

The overall mass balance at junction C is

$$\text{input} = \text{output}$$

$$\begin{aligned} E_o + B &= P \\ E_o &= P - B = 285.7\ \frac{\text{lbm}}{\text{hr}} - B \end{aligned}$$

The mass balance for detergent at junction C is

$$\begin{aligned} &E_o\left(0.58\ \frac{\text{lbm } D}{\text{lbm}}\right) + B\left(0.12\ \frac{\text{lbm } D}{\text{lbm}}\right) \\ &\quad = P\left(0.42\ \frac{\text{lbm } D}{\text{lbm}}\right) \\ &\left(285.7\ \frac{\text{lbm}}{\text{hr}} - B\right)\left(0.58\ \frac{\text{lbm } D}{\text{lbm}}\right) \\ &\qquad + B\left(0.12\ \frac{\text{lbm } D}{\text{lbm}}\right) \\ &\quad = \left(285.7\ \frac{\text{lbm}}{\text{hr}}\right)\left(0.42\ \frac{\text{lbm } D}{\text{lbm}}\right) \\ &165.7\ \frac{\text{lbm } D}{\text{hr}} - 0.58B\ \frac{\text{lbm } D}{\text{hr}} + 0.12B\ \frac{\text{lbm } D}{\text{hr}} \\ &\quad = 120.0\ \text{lbm } D/\text{hr} \\ &B = 99.35\ \text{lbm/hr} \end{aligned}$$

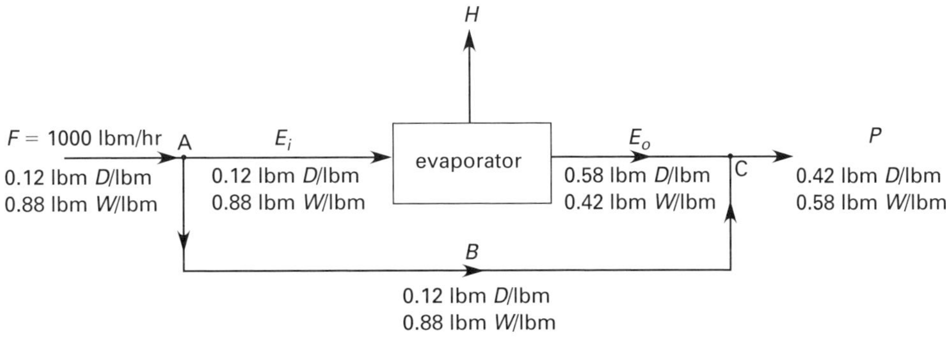

Illustration for Solution 1.1

The fraction of the feed that bypasses the evaporator is

$$\frac{B}{F} = \frac{99.35 \ \frac{\text{lbm}}{\text{hr}}}{1000 \ \frac{\text{lbm}}{\text{hr}}} = \boxed{0.0993}$$

The answer is A.

1.4. An overall mass balance at junction A results in

$$F = E_i + B$$

$$E_i = F - B = 1000 \ \frac{\text{lbm}}{\text{hr}} - 99.35 \ \frac{\text{lbm}}{\text{hr}}$$

$$= \boxed{900.65 \text{ lbm/hr}}$$

The answer is D.

1.5. Without the bypass stream, the stream coming out of the evaporator will have a concentration of 58% detergent and 42% water. The schematic for this process is

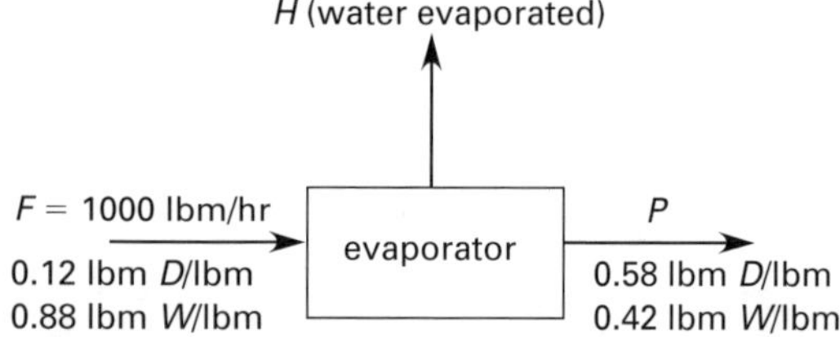

The mass balance for the detergent is

$$\left(1000 \ \frac{\text{lbm}}{\text{hr}}\right)\left(0.12 \ \frac{\text{lbm } D}{\text{lbm}}\right) = P\left(0.58 \ \frac{\text{lbm } D}{\text{lbm}}\right)$$

$$P = 206.9 \text{ lbm/hr}$$

An overall mass balance results in the following equation.

$$F = P + H$$

Therefore,

$$H = F - P = 1000 \ \frac{\text{lbm}}{\text{hr}} - 206.9 \ \frac{\text{lbm}}{\text{hr}}$$

$$= \boxed{793.1 \text{ lbm/hr}}$$

The answer is B.

SOLUTION 2

2.1. The moles of oxygen and nitrogen in the air supplied are

$$\left(500 \ \frac{\text{mol air}}{\text{h}}\right)\left(0.21 \ \frac{\text{mol } O_2}{\text{mol air}}\right) = 105 \text{ mol } O_2/\text{h}$$

$$500 \ \frac{\text{mol air}}{\text{h}} - 105 \ \frac{\text{mol } O_2}{\text{h}} = \boxed{395 \text{ mol } N_2/\text{h}}$$

The answer is D.

2.2. Not all of the methane participates in the two reactions, and it is clear that excess air is being supplied. So the product stream includes unreacted methane and oxygen. The product stream also includes nitrogen, which is inert.

A schematic for the process is drawn as follows.

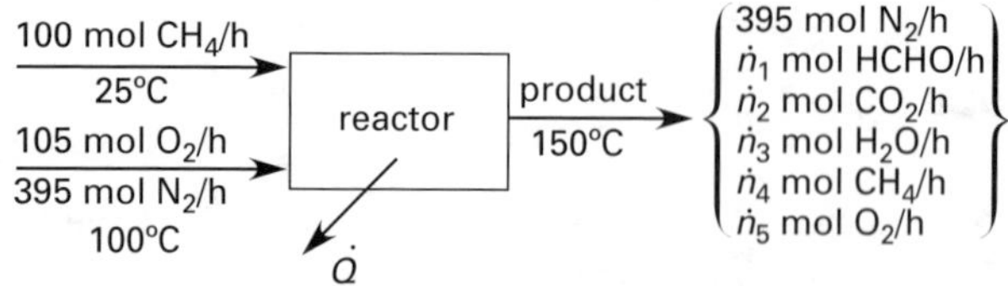

30% conversion to formaldehyde implies that 30 mol/h methane is consumed in reaction 1. 20% of the remaining methane participates in reaction 2. The methane consumed in reaction 2 is

$$(0.20)\left(100 \ \frac{\text{mol}}{\text{h}} - 30 \ \frac{\text{mol}}{\text{h}}\right) = 14 \text{ mol/h}$$

The molar flow rate of methane in the product stream, $\dot{n}_4$, is calculated using the material balance equation.

$$\text{input} + \text{generation} = \text{output} + \text{consumption}$$

For CH_4,

$$100 \ \frac{\text{mol}}{\text{h}} + 0 \ \frac{\text{mol}}{\text{h}} = \dot{n}_4 + \left(30 \ \frac{\text{mol}}{\text{h}} + 14 \ \frac{\text{mol}}{\text{h}}\right)$$

$$\dot{n}_4 = \boxed{56 \text{ mol } CH_4/\text{h}}$$

The answer is B.

2.3. The material balance equation is applied to H_2O.

$$0 \ \frac{\text{mol}}{\text{h}} + \left(30 \ \frac{\text{mol}}{\text{h}} + \left(2 \ \frac{\text{mol}}{\text{h}}\right)\left(14 \ \frac{\text{mol}}{\text{h}}\right)\right)$$

$$= \dot{n}_3 + 0 \ \frac{\text{mol}}{\text{h}}$$

$$\dot{n}_3 = \boxed{58 \text{ mol } H_2O/\text{h}}$$

The answer is D.

2.4. This is a reactive process with two competing reactions. The enthalpies of each reactant and product species are calculated using elements that constitute each species at 25°C as the reference state. That is, C(s), O_2, H_2, and N_2 at 25°C are used as references in calculating enthalpies of reactants and products. The enthalpies are zero at the reference state. The following data are used in enthalpy calculations: standard heats of formation, heat capacity data, and mean heat capacity data. The molar enthalpy of oxygen going into the reactor is

$$\begin{aligned}\left(\hat{\mathrm{H}}_{\mathrm{O}_2}\right)_{\mathrm{in}} &= \left(\hat{\mathrm{H}}_{\mathrm{O}_2}\right)_{100^\circ\mathrm{C}} \\ &= \left(\overline{C}_{P_{\mathrm{O}_2}}\right)_{100^\circ\mathrm{C}}(100^\circ\mathrm{C} - 25^\circ\mathrm{C}) \\ &= \left(29.80\ \frac{\mathrm{J}}{\mathrm{mol\cdot^\circ C}}\right)(75^\circ\mathrm{C})\left(\frac{1\ \mathrm{kJ}}{1000\ \mathrm{J}}\right) \\ &= \boxed{2.235\ \mathrm{kJ/mol}}\end{aligned}$$

The answer is C.

2.5. The molar enthalpy of formaldehyde coming out of the reactor is calculated using the heat of formation and specific heat data.

$$\begin{aligned}\left(\hat{H}_{\mathrm{HCHO}}\right)_{\mathrm{out}} &= \left(\hat{H}_{\mathrm{HCHO}}\right)_{150^\circ\mathrm{C}} \\ &= \left(\Delta\hat{H}_f\right)_{\mathrm{HCHO}} + \int_{25^\circ\mathrm{C}}^{150^\circ\mathrm{C}} (C_p)_{\mathrm{HCHO}}\,dT \\ \left(\Delta\hat{H}_f\right)_{\mathrm{HCHO}} &= -115.9\ \mathrm{kJ/mol} \\ (C_p)_{\mathrm{HCHO}} &= 34.28 + 4.628\times10^{-2}\ T\end{aligned}$$

Note that C_p is in J/mol·°C and T is in °C.

$$\begin{aligned}\left(\hat{H}_{\mathrm{HCHO}}\right)_{\mathrm{out}} &= -115.9\ \frac{\mathrm{kJ}}{\mathrm{mol}} + \frac{1\ \mathrm{kJ}}{1000\ \mathrm{J}}\int_{25^\circ\mathrm{C}}^{150^\circ\mathrm{C}} \\ &\quad\times\left(34.28 + 4.628\times10^{-2}\ T\right)dT \\ &= -115.9\ \frac{\mathrm{kJ}}{\mathrm{mol}} + \left(\frac{1\ \mathrm{kJ}}{1000\ \mathrm{J}}\right) \\ &\quad\times\left(\begin{array}{l}(34.28)(150^\circ\mathrm{C} - 25^\circ\mathrm{C}) \\ \quad + \left(\dfrac{4.628\times10^{-2}}{2}\right) \\ \quad \times\left((150^\circ\mathrm{C})^2 - (25^\circ\mathrm{C})^2\right)\end{array}\right) \\ &= -115.9\ \frac{\mathrm{kJ}}{\mathrm{mol}} + 4.79\ \frac{\mathrm{kJ}}{\mathrm{mol}} \\ &= \boxed{-111.1\ \mathrm{kJ/mol}}\end{aligned}$$

The answer is D.

2.6. The molar enthalpy of water vapor coming out of the reactor is calculated using the heat of formation and mean heat capacity data. Note that the heat of formation is that of H_2O (vapor).

$$\begin{aligned}\left(\hat{H}_{\mathrm{H_2O}}\right)_{\mathrm{out}} &= \left(\hat{H}_{\mathrm{H_2O}}\right)_{150^\circ\mathrm{C}} \\ &= \left(\Delta\hat{H}_f\right)_{\mathrm{H_2O}} \\ &\quad + \left(\overline{C}_{p_{\mathrm{H_2O}}}\right)_{150^\circ\mathrm{C}}(150^\circ\mathrm{C} - 25^\circ\mathrm{C}) \\ &= -241.8\ \frac{\mathrm{kJ}}{\mathrm{mol}} \\ &\quad + \left(\frac{34.13\ \dfrac{\mathrm{J}}{\mathrm{mol\cdot^\circ C}}}{1000\ \dfrac{\mathrm{J}}{\mathrm{kJ}}}\right)(125^\circ\mathrm{C}) \\ &= \boxed{-237.5\ \mathrm{kJ/mol}}\end{aligned}$$

The answer is D.

2.7. To obtain the rate of heat removal from the reactor, determine the unknown molar flow rates, $\dot{n}_1$, $\dot{n}_2$ and $\dot{n}_5$, by applying the material balance equation for HCHO, O_2, and CO_2, as follows.

$$\text{input} + \text{generation} = \text{output} + \text{consumption}$$

For HCHO,

$$0\ \frac{\mathrm{mol\ HCHO}}{\mathrm{h}} + 30\ \frac{\mathrm{mol\ HCHO}}{\mathrm{h}} = \dot{n}_1 + 0\ \frac{\mathrm{mol\ HCHO}}{\mathrm{h}}$$
$$\dot{n}_1 = 30\ \mathrm{mol\ HCHO/h}$$

For O_2,

$$105\ \frac{\mathrm{mol\ O_2}}{\mathrm{h}} + 0\ \frac{\mathrm{mol\ O_2}}{\mathrm{h}} = \dot{n}_5 + \left(\begin{array}{l}30\ \dfrac{\mathrm{mol\ O_2}}{\mathrm{h}} + (2) \\ \quad\times\left(14\ \dfrac{\mathrm{mol\ O_2}}{\mathrm{h}}\right)\end{array}\right)$$
$$\dot{n}_5 = 47\ \mathrm{mol\ O_2/h}$$

For CO_2,

$$0\ \frac{\mathrm{mol\ CO_2}}{\mathrm{h}} + 14\ \frac{\mathrm{mol\ CO_2}}{\mathrm{h}} = \dot{n}_2 + 0\ \frac{\mathrm{mol\ CO_2}}{\mathrm{h}}$$
$$\dot{n}_2 = 14\ \mathrm{mol\ CO_2/h}$$

The unknown molar enthalpies of the reactant and product species are also calculated as shown. The standard heat of formation of CH_4 is

$$\left(\hat{H}_{\mathrm{CH_4}}\right)_{\mathrm{in}} = \left(\hat{H}_{\mathrm{CH_4}}\right)_{25^\circ\mathrm{C}} = (\hat{H}_f)_{\mathrm{CH_4}} = -74.85\ \mathrm{kJ/mol}$$

$$\begin{aligned}\left(\hat{H}_{\mathrm{N_2}}\right)_{\mathrm{in}} &= \left(\hat{H}_{\mathrm{N_2}}\right)_{100^\circ\mathrm{C}} \\ &= \left(\overline{C}_{P_{\mathrm{N_2}}}\right)_{100^\circ\mathrm{C}}(100^\circ\mathrm{C} - 25^\circ\mathrm{C}) \\ &= \left(29.16\ \frac{\mathrm{J}}{\mathrm{mol\cdot^\circ C}}\right)(75^\circ\mathrm{C})\left(\frac{1\ \mathrm{kJ}}{1000\ \mathrm{J}}\right) \\ &= 2.187\ \mathrm{kJ/mol}\end{aligned}$$

$$\left(\hat{H}_{N_2}\right)_{out} = \left(\hat{H}_{N_2}\right)_{150°C}$$
$$= \left(\overline{C}_{P_{N_2}}\right)_{150°C}(150°C - 25°C)$$
$$= \left(29.24\ \frac{J}{mol\cdot°C}\right)(125°C)\left(\frac{1\ kJ}{1000\ J}\right)$$
$$= 3.655\ kJ/mol$$

$$\left(\hat{H}_{CH_4}\right)_{out} = \left(\hat{H}_{CH_4}\right)_{150°C}$$
$$= \left(\Delta\hat{H}_f\right)_{CH_4} + \int_{25°C}^{150°C}(C_p)_{CH_4}\,dT$$
$$\left(\Delta\hat{H}_f\right)_{CH_4} = -74.85\ kJ/mol$$

The molar specific heat can be estimated from the empirical correlation.

$$(C_p)_{CH_4} = 34.31 + 5.469\times10^{-2}T$$

Note that C_p is in J/mol·C and T is in °C.

$$\left(\hat{H}_{CH_4}\right)_{out} = -74.85\ \frac{kJ}{mol} + \frac{1}{1000}\int_{25°C}^{150°C}\times\left(34.31 + 5.469\times10^{-2}T\right)dT$$
$$= -74.85\ \frac{kJ}{mol} + \left(\frac{1\ kJ}{1000\ J}\right)\times\left(\begin{array}{c}(34.31)(150°C - 25°C)\\ +\left(\dfrac{5.469\times10^{-2}}{2}\right)\\ \times\left((150°C)^2 - (25°C)^2\right)\end{array}\right)$$
$$= -74.85\ \frac{kJ}{mol} + 4.87\ \frac{kJ}{mol}$$
$$= -69.98\ kJ/mol$$

$$\left(\hat{H}_{CO_2}\right)_{out} = \left(\hat{H}_{CO_2}\right)_{150°C}$$
$$= (\Delta H_f)CO_2 + \left(\overline{C}_{p_{CO_2}}\right)_{150°C}(150°C - 25°C)$$
$$= -393.5\ \frac{kJ}{mol} + \left(\frac{39.04\ kJ}{1000\ mol\cdot°C}\right)(125°C)$$
$$= -388.6\ kJ/mol$$

$$\left(\hat{H}_{O_2}\right)_{out} = \left(\hat{H}_{O_2}\right)_{150°C}$$
$$= \left(\overline{C}_{P_{O_2}}\right)_{150°C}(150°C - 25°C)$$
$$= \left(30.06\ \frac{J}{mol\cdot°C}\right)(125°C)\left(\frac{1\ kJ}{1000\ J}\right)$$
$$= 3.758\ kJ/mol$$

An enthalpy table listing the $\dot{n}_i$ and $\hat{H}_i$ for each reactant and product is constructed using calculated results. $\dot{n}_i$ is in mol/h, and $\hat{H}_i$ is in kJ/mol.

species	$\dot{n}_{in}$	$\hat{H}_{in}$	$\dot{n}_{out}$	$\hat{H}_{out}$
N_2	395	2.187	395	3.655
O_2	105	2.235	47	3.758
CH_4	100	−74.85	56	−69.98
HCHO	–	–	30	−111.1
CO_2	–	–	14	−388.6
H_2O	–	–	58	−237.5

$$\sum_{inlet}\dot{n}_i\hat{H}_i = -6386\ kJ/h \qquad \sum_{outlet}\dot{n}_i\hat{H}_i = -24\,847\ kJ/h$$

The energy balance at steady state is

energy in = energy out [refer to the schematic]

$$\sum_{inlet}\dot{n}_i\hat{H}_i = \sum_{outlet}\dot{n}_iH_i + \dot{Q}$$
$$\dot{Q} = \sum_{inlet}\dot{n}_i\hat{H}_i - \sum_{outlet}\dot{n}_i\hat{H}_i$$
$$= -6386\ \frac{kJ}{h} - \left(-24\,847\ \frac{kJ}{h}\right)$$
$$= 18\,461\ kJ/h$$
$$\dot{Q} = \left(18\,461\ \frac{kJ}{h}\right)\left(1000\ \frac{J}{kJ}\right)\times\left(9.486\times10^{-4}\ \frac{Btu}{J}\right)$$
$$= 17\,512\ Btu/h$$

The rate of heat removal from the reactor is

17 512 Btu/h.

The answer is C.

SOLUTION 3

3.1. Basis: Assume 100 kmol of dry combustion products.

The reaction equation is written based on the Orsat analysis of combustion products. The Orsat analysis does not include water vapor, but water vapor is a combustion product.

$$C_aH_b + cO_2 + dN_2 \longrightarrow 13CO_2 + 0.5CO + 3.5O_2 + 83N_2 + eH_2O$$

Balances are written for each species as shown.

The C balance is

$$a = 13 + 0.5 = 13.5$$

The N_2 balance is

$$d = 83$$

The N_2/O_2 ratio in air is

$$\frac{d}{c} = 3.76$$
$$c = \frac{d}{3.76} = \frac{83}{3.76} = 22.07$$

The O balance is

$$2c = 26 + 0.5 + 7.0 + e$$
$$e = 10.64$$

10.64 mol of water vapor are formed.

The answer is B.

3.2. The H balance is

$$b = 2e = (2)(10.64) = 21.28$$

Therefore, the molecular formula of the fuel is

$C_{13.5}H_{21.28}$.

The answer is C.

3.3.

$$\begin{aligned}(r_{a/f})_{\text{molar}} &= \frac{n_{\text{air}}}{n_{\text{fuel}}} = \frac{n_{O_2} + n_{N_2}}{n_{\text{fuel}}} \\ &= \frac{22.07 \text{ kmol } O_2 + 83 \text{ kmol } N_2}{1 \text{ kmol fuel}} \\ &= 105.07 \text{ kmol air/kmol fuel}\end{aligned}$$

$$\begin{aligned}(r_{a/f})_{\text{mass}} &= \frac{m_{\text{air}}}{m_{\text{fuel}}} \\ &= \frac{(105.07 \text{ kmol air})\left(29 \ \frac{\text{kg}}{\text{kmol}}\right)}{(1 \text{ kmol fuel})\left(\begin{array}{c}(12)\left(13.5 \ \frac{\text{kg}}{\text{kmol}}\right) \\ + (21.28)\left(1 \ \frac{\text{kg}}{\text{kmol}}\right)\end{array}\right)} \\ &= \boxed{16.63 \text{ kg air/kg fuel}}\end{aligned}$$

The answer is D.

3.4. The theoretical combustion equation is obtained by balancing the following equation.

$$C_{13.5}H_{21.28} + aO_2 + bN_2 \longrightarrow cCO_2 + dH_2O + eN_2$$

The C balance is

$$c = 13.5$$

The H balance is

$$21.28 = 2d$$
$$d = 10.64$$

The O balance is

$$2a = 2c + d = (2)(13.5) + 10.64 = 37.64$$
$$a = 18.82$$

The N_2/O_2 ratio is

$$b = 3.76a = (3.76)(18.82) = 70.76$$

The N_2 balance is

$$e = b = 70.76$$

The theoretical combustion equation is

$$\begin{aligned}&C_{13.5}H_{21.28} + 18.82O_2 + 70.76N_2 \longrightarrow \\ &\quad 13.5CO_2 + 10.64H_2O + 70.76N_2\end{aligned}$$

The theoretical air required is

$$18.82 \text{ kmol } O_2 + 70.76 \text{ kmol } N_2 = 89.58 \text{ kmol air}$$

The actual number of moles of air supplied is 105.07 kmol air (from the solution to Prob. 3.3).

The percent excess air supplied is

$$\left(\frac{105.07 \text{ kmol} - 89.58 \text{ kmol}}{89.58 \text{ kmol}}\right) \times 100\% = \boxed{17.29\%}$$

The answer is A.

3.5. From the actual combustion equation, the total number of moles of combustion products is

$$100 \text{ kmol} + 10.64 \text{ kmol} = 110.64 \text{ kmol}$$
$$\text{mol of water vapor formed} = 10.64 \text{ kmol}$$

The mole fraction of water vapor is

$$\frac{10.64 \text{ kmol}}{110.64 \text{ kmol}} = 0.0962$$

Since the mole fraction is equal to the pressure fraction,

$$\frac{p_{H_2O}}{p} = 0.0962$$

$$\begin{aligned} p_{H_2O} &= 0.0962p \\ &= (0.0962)(14.7 \text{ psi}) \\ &= 1.414 \text{ psi} \end{aligned}$$

From steam tables, the saturation temperature at 1.414 psi is 112°F. Therefore, the dew point of the combustion product is $\boxed{112°\text{F.}}$

The answer is A.

SOLUTION 4

4.1. A schematic diagram for the process is shown.

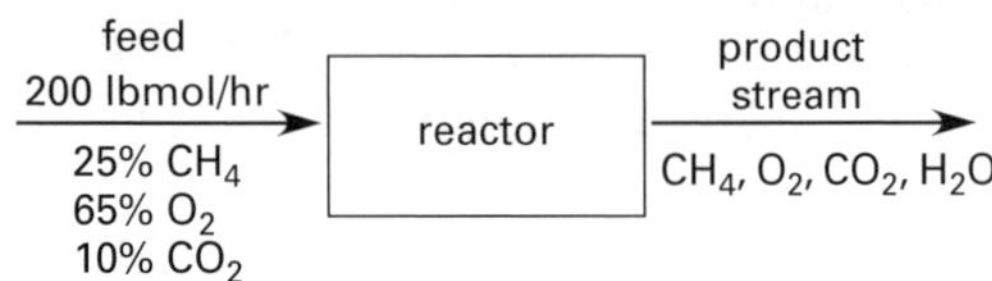

The stoichiometric reaction equation is

$$CH_4 + 2O_2 \longrightarrow CO_2 + 2H_2O$$

The molar flow rate of each component in the feed stream is calculated as shown.

The molar flow rate of CH_4 in the feed stream is

$$\left(0.25 \ \frac{\text{lbmol } CH_4}{\text{lbmol}}\right)\left(200 \ \frac{\text{lbmol}}{\text{hr}}\right) = 50 \text{ lbmol } CH_4/\text{hr}$$

The molar flow rate of O_2 in the feed stream is

$$\left(0.65 \ \frac{\text{lbmol } O_2}{\text{lbmol}}\right)\left(200 \ \frac{\text{lbmol}}{\text{hr}}\right) = 130 \text{ lbmol } O_2/\text{hr}$$

The molar flow rate of CO_2 in the feed stream is

$$\left(0.10 \ \frac{\text{lbmol } CO_2}{\text{lbmol}}\right)\left(200 \ \frac{\text{lbmol}}{\text{hr}}\right) = 20 \text{ lbmol } CO_2/\text{hr}$$

For 50 lbmol/hr of methane, the stoichiometric requirement of oxygen is 100 lbmol/hr. Since 130 lbmol/hr of oxygen is supplied, oxygen is the excess reactant and methane is the limiting reactant. The percent excess O_2 supplied is

$$\left(\frac{130 \ \frac{\text{lbmol}}{\text{hr}} - 100 \ \frac{\text{lbmol}}{\text{hr}}}{100 \ \frac{\text{lbmol}}{\text{hr}}}\right) \times 100\% = \boxed{30\%}$$

The answer is C.

4.2. The mole balance for CH_4 is as follows.

$$\text{input} + \text{generation} = \text{output} + \text{consumption}$$

$$50 \ \frac{\text{lbmol}}{\text{hr}} + 0 \ \frac{\text{lbmol}}{\text{hr}} = \text{output} + (0.9)\left(50 \ \frac{\text{lbmol}}{\text{hr}}\right)$$

The output of CH_4 is

$$\left(50 \ \frac{\text{lbmol}}{\text{hr}}\right)(1 - 0.9) = \boxed{5 \text{ lbmol/hr}}$$

The answer is B.

4.3. The mole balance for O_2 is as follows.

$$\text{input} + \text{generation} = \text{output} + \text{consumption}$$

$$130 \ \frac{\text{lbmol}}{\text{hr}} + 0 \ \frac{\text{lbmol}}{\text{hr}} = \text{output} + (2)(0.9)\left(50 \ \frac{\text{lbmol}}{\text{hr}}\right)$$

The output of O_2 is

$$130 \ \frac{\text{lbmol}}{\text{hr}} - (2)(0.9)\left(50 \ \frac{\text{lbmol}}{\text{hr}}\right) = 40 \text{ lbmol/hr}$$

The mole balance for CO_2 is as follows.

$$\text{input} + \text{generation} = \text{output} + \text{consumption}$$

$$20 \ \frac{\text{lbmol}}{\text{hr}} + (0.9)\left(50 \ \frac{\text{lbmol}}{\text{hr}}\right) = \text{output} + 0 \ \frac{\text{lbmol}}{\text{hr}}$$

The output of CO_2 is

$$20 \ \frac{\text{lbmol}}{\text{hr}} + (0.9)\left(50 \ \frac{\text{lbmol}}{\text{hr}}\right) = 65 \text{ lbmol/hr}$$

The mole balance for H_2O is as follows.

$$\text{input} + \text{generation} = \text{output} + \text{consumption}$$

$$0 \ \frac{\text{lbmol}}{\text{hr}} + (2)(0.9)\left(50 \ \frac{\text{lbmol}}{\text{hr}}\right) = \text{output} + 0 \ \frac{\text{lbmol}}{\text{hr}}$$

The output of H_2O is

$$(2)(0.9)\left(50 \ \frac{\text{lbmol}}{\text{hr}}\right) = 90 \text{ lbmol/hr}$$

The molar flow rate of product gas is

$$\begin{aligned} &5 \ \frac{\text{lbmol } CH_4}{\text{hr}} + 40 \ \frac{\text{lbmol } O_2}{\text{hr}} \\ &\quad + 65 \ \frac{\text{lbmol } CO_2}{\text{hr}} + 90 \ \frac{\text{lbmol } H_2O}{\text{hr}} = \boxed{200 \text{ lbmol/hr}} \end{aligned}$$

The answer is B.

4.4. The mole fraction of carbon dioxide in the product stream is

$$y_{CO_2} = \frac{n_{CO_2}}{n} = \frac{65\ \frac{\text{lbmol}}{\text{hr}}}{200\ \frac{\text{lbmol}}{\text{hr}}} = \boxed{0.325}$$

The answer is D.

4.5. The mole fraction of oxygen in the product stream is

$$y_{O_2} = \frac{n_{O_2}}{n} = \frac{40\ \frac{\text{lbmol}}{\text{hr}}}{200\ \frac{\text{lbmol}}{\text{hr}}} = 0.20$$

The partial pressure of oxygen is

$$p_{O_2} = p y_{O_2} = \left(14.7\ \frac{\text{lbf}}{\text{in}^2}\right)(0.2) = \boxed{2.94\ \text{lbf/in}^2}$$

The answer is A.

4.6.

$$T = 360°\text{F} + 460° = 820°\text{R}$$

$$p\dot{V} = \dot{n}\overline{R}T$$

$$\dot{V} = \frac{\dot{n}\overline{R}T}{p} = \left(\frac{\left(200\ \frac{\text{lbmol}}{\text{hr}}\right)\left(1545\ \frac{\text{ft-lbf}}{\text{lbmol-}°\text{R}}\right)(820°\text{R})}{\left(14.7\ \frac{\text{lbf}}{\text{in}^2}\right)\left(144\ \frac{\text{in}^2}{\text{ft}^2}\right)}\right) \times \left(\frac{1\ \text{hr}}{60\ \text{min}}\right) = \boxed{1995\ \text{ft}^3/\text{min}}$$

The answer is C.

Thermodynamics

PROBLEM 1

Carbon dioxide dissociates into carbon monoxide and oxygen at a temperature of 3500°C and a pressure of 1 atm. Under these conditions, the equilibrium constant, K_p, is 0.31.

1.1. What is the equilibrium conversion of CO_2?

(A) 24%
(B) 31%
(C) 43%
(D) 58%

1.2. What is the partial pressure of O_2 at equilibrium?

(A) 0.11 atm
(B) 0.18 atm
(C) 0.24 atm
(D) 0.35 atm

1.3. If 25 mol of CO_2 are present at the beginning of the reaction, what is the total number of moles at equilibrium?

(A) 28 mol
(B) 30 mol
(C) 36 mol
(D) 39 mol

PROBLEM 2

A liquid mixture containing benzene and toluene is in equilibrium with the vapor phase at a temperature of 70°F. The composition of the liquid is 40% benzene by mass.

2.1. What is the mole fraction of benzene in the liquid phase?

(A) 0.38
(B) 0.44
(C) 0.51
(D) 0.62

2.2. What is the total pressure in the vapor phase?

(A) 0.061 atm
(B) 0.072 atm
(C) 0.083 atm
(D) 0.094 atm

2.3. What is the mole fraction of benzene in the vapor phase?

(A) 0.51
(B) 0.58
(C) 0.65
(D) 0.73

PROBLEM 3

A refrigerator operating on the reversed Carnot cycle maintains the temperature of a cold storage tank at −20°F. The surrounding temperature can reach a maximum value of 90°F.

3.1. What is the coefficient of performance (COP) for this refrigerator?

(A) 2.0
(B) 3.0
(C) 4.0
(D) 5.0

3.2. The system uses a 6 hp (electrical power input) motor with 80% efficiency. What is the refrigeration in tons?

(A) 1.3 tons
(B) 3.8 tons
(C) 4.1 tons
(D) 6.4 tons

3.3. The refrigeration cycle rejects heat to ambient air. What is the heat transfer rate to ambient air?

(A) 17 Btu/sec
(B) 25 Btu/sec
(C) 33 Btu/sec
(D) 37 Btu/sec

3.4. The temperature of the cold storage tank needs to be maintained at −40°F instead of at −20°F. The same motor as in Prob. 3.2 is used. What is the refrigeration in tons?

(A) 2.6 tons
(B) 3.3 tons
(C) 4.1 tons
(D) 5.2 tons

PROBLEM 4

6 lbm of ethane is at 14.7 psia and 70°F. The ethane is heated at constant volume to a final temperature of 180°F. At 80°F, for ethane, $c_p = 0.427$ Btu/lbm-°R and $c_v = 0.361$ Btu/lbm-°R.

4.1. What is the change in internal energy?

(A) 190 Btu
(B) 240 Btu
(C) 280 Btu
(D) 320 Btu

4.2. What is the change in enthalpy?

(A) 190 Btu
(B) 240 Btu
(C) 280 Btu
(D) 320 Btu

4.3. What is the change in entropy?

(A) 0.41 Btu/°R
(B) 0.68 Btu/°R
(C) 0.82 Btu/°R
(D) 0.97 Btu/°R

4.4. What is the work done?

(A) −1.2 Btu
(B) 0.0 Btu
(C) 1.2 Btu
(D) 2.1 Btu

4.5. What is the heat added?

(A) 240 Btu
(B) 280 Btu
(C) 330 Btu
(D) 380 Btu

For Probs. 4.6–4.8, assume the gas is heated at constant pressure instead of constant volume.

4.6. What is the heat added?

(A) 240 Btu
(B) 280 Btu
(C) 320 Btu
(D) 370 Btu

4.7. What is the work done?

(A) 0 Btu
(B) 22 Btu
(C) 36 Btu
(D) 43 Btu

4.8. What is the change in entropy?

(A) 0.41 Btu/°R
(B) 0.48 Btu/°R
(C) 0.57 Btu/°R
(D) 0.71 Btu/°R

For Probs. 4.9 and 4.10, use the following expression for the molar heat capacity of ethane.

$$\bar{c}_p = 1.648 + \left(2.291 \times 10^{-2}\right) T - \left(0.4722 \times 10^{-5}\right) T^2$$

$\bar{c}_p$ is in Btu/lbmol-°R, and T is in °R.

4.9. Using the preceding expression for $\bar{c}_p$, the percent error in the change in enthalpy determined in Prob. 4.2 is

(A) 1.3%
(B) 2.6%
(C) 4.6%
(D) 5.3%

4.10. If ethane is heated from 70°F to 700°F, the mean heat capacity, $\bar{c}_{p_{\text{mean}}}$, in this range is

(A) 0.43 Btu/lbm-°R
(B) 0.51 Btu/lbm-°R
(C) 0.58 Btu/lbm-°R
(D) 0.63 Btu/lbm-°R

Thermodynamics Solutions

SOLUTION 1

1.1. The reaction is

$$CO_2(g) \rightleftharpoons CO(g) + \tfrac{1}{2}O_2(g)$$

Start with 1 mol of CO_2. If x (in mol) is the equilibrium conversion of CO_2, then at equilibrium,

$$\text{no. of moles of } CO_2 = 1 - x$$
$$\text{no. of moles of } CO = x$$
$$\text{no. of moles of } O_2 = \frac{x}{2}$$

The total number of moles at equilibrium is

$$(1-x) + x + \frac{x}{2} = \frac{2+x}{2}$$

Since this is a gas-phase reaction, the pressure fraction equals the mole fraction.

$$\frac{p_{CO_2}}{p} = \frac{1-x}{\frac{2+x}{2}} = \frac{2(1-x)}{2+x}$$

$$p_{CO_2} = \left(\frac{2(1-x)}{2+x}\right)p$$

$$\frac{p_{CO}}{p} = \frac{x}{\frac{2+x}{2}} = \frac{2x}{2+x}$$

$$p_{CO} = \left(\frac{2x}{2+x}\right)p$$

$$\frac{p_{O_2}}{p} = \frac{\frac{x}{2}}{\frac{2+x}{2}} = \frac{x}{2+x}$$

$$p_{O_2} = \left(\frac{x}{2+x}\right)p$$

The equilibrium constant K_p is

$$K_p = \frac{p^1_{CO}p^{0.5}_{O_2}}{p^1_{CO_2}} = \frac{\left(\frac{2x}{2+x}\right)p\sqrt{\frac{x}{2+x}}\sqrt{p}}{\left(\frac{(2)(1-x)}{2+x}\right)p}$$

$$K_p = \frac{x^{\frac{3}{2}}\sqrt{p}}{\sqrt{2+x}\,(1-x)}$$

Since $p = 1$ atm,

$$K_p = \frac{x^{\frac{3}{2}}}{\sqrt{2+x}\,(1-x)}$$

Since $K_p = 0.31$,

$$\frac{x^{\frac{3}{2}}}{\sqrt{2+x}\,(1-x)} = 0.31$$

By trial and error, $x = 0.425$.

Therefore, the equilibrium conversion of CO_2 is $\boxed{42.5\%.}$

The answer is C.

1.2. The partial pressure of O_2 at equilibrium is given by

$$\begin{aligned} p_{O_2} &= \left(\frac{x}{2+x}\right)p \\ &= \left(\frac{0.425}{2+0.425}\right)(1 \text{ atm}) \\ &= \boxed{0.1753 \text{ atm}} \end{aligned}$$

The answer is B.

1.3. With 42.5% equilibrium conversion of CO_2,

$$\begin{aligned} \text{mol of } CO_2 \text{ reacted} &= (0.425)(25 \text{ mol}) \\ &= 10.63 \text{ mol} \\ \text{mol of CO formed} &= 10.63 \text{ mol} \\ \text{mol of } O_2 \text{ formed} &= \frac{10.63 \text{ mol}}{2} = 5.32 \text{ mol} \\ \text{mol of } CO_2 \text{ at equilibrium} &= 25 \text{ mol} - 10.63 \text{ mol} \\ &= 14.37 \text{ mol} \end{aligned}$$

The total number of moles at equilibrium is

$$10.63 \text{ mol} + 5.32 \text{ mol} + 14.37 \text{ mol} = \boxed{30.32 \text{ mol}}$$

The answer is B.

SOLUTION 2

2.1. The molecular weight of benzene (C_6H_6) is 78 g/mol.

The molecular weight of toluene ($C_6H_5CH_3$) is 92 g/mol.

Using 100 g of liquid mixture as the basis,

$$\text{mol of benzene} = \frac{m_b}{M_b} = \frac{40 \text{ g}}{78 \frac{\text{g}}{\text{mol}}} = 0.5128 \text{ mol}$$

$$\text{mol of toluene} = \frac{m_t}{M_t} = \frac{60 \text{ g}}{92 \frac{\text{g}}{\text{mol}}} = 0.6522 \text{ mol}$$

The total number of moles is

$$\begin{aligned} n &= 0.5128 \text{ mol} + 0.6522 \text{ mol} \\ &= 1.165 \text{ mol} \end{aligned}$$

The mole fraction of benzene in the liquid phase is

$$x_b = \left(\frac{n_b}{n}\right) = \frac{0.5128 \text{ mol}}{1.165 \text{ mol}} = \boxed{0.44}$$

The answer is B.

2.2. At 70°F, the vapor pressures of benzene and toluene are

$$\begin{aligned} P_{\text{vap},b} &= 0.1000 \text{ atm} \\ P_{\text{vap},t} &= 0.0300 \text{ atm} \end{aligned}$$

Using Raoult's law, the partial pressures of benzene and toluene in the vapor phase can be calculated as shown.

$$\begin{aligned} p_b &= x_b P_{\text{vap},b} = (0.44)(0.1000 \text{ atm}) = 0.044 \text{ atm} \\ p_t &= x_t P_{\text{vap},t} = (0.56)(0.0300 \text{ atm}) = 0.0168 \text{ atm} \end{aligned}$$

The total pressure in the vapor phase is the sum of the partial pressures.

$$p = p_b + p_t = 0.044 \text{ atm} + 0.0168 \text{ atm} = \boxed{0.0608 \text{ atm}}$$

The answer is A.

2.3. In the vapor phase, the mole fraction of benzene is

$$y_b = \frac{p_b}{p} = \frac{0.044 \text{ atm}}{0.0608 \text{ atm}} = \boxed{0.73}$$

The answer is D.

SOLUTION 3

3.1.

$$\begin{aligned} T_H &= 90°\text{F} + 460° = 550°\text{R} \\ T_L &= -20°\text{F} + 460° = 440°\text{R} \\ \text{COP} &= \frac{T_L}{T_H - T_L} = \frac{440°\text{R}}{550°\text{R} - 440°\text{R}} \\ &= \boxed{4.0} \end{aligned}$$

The answer is C.

3.2.

$$\begin{aligned} \text{COP} &= \frac{\dot{Q}_L}{\dot{W}} \\ \dot{Q}_L &= (\text{COP})\dot{W} \end{aligned}$$

$\dot{Q}_L$ is the rate of heat removal from the refrigerated space; $\dot{W}$ is the power supplied.

$$\dot{W} = (0.8)(6 \text{ hp})\left(\frac{0.7068 \frac{\text{Btu}}{\text{sec}}}{1 \text{ hp}}\right) = 3.393 \text{ Btu/sec}$$

$$\begin{aligned} \dot{Q}_L &= (\text{COP})\dot{W} = (4)\left(3.393 \frac{\text{Btu}}{\text{sec}}\right) \\ &= \left(13.6 \frac{\text{Btu}}{\text{sec}}\right)\left(\frac{1 \text{ ton}}{3.333 \frac{\text{Btu}}{\text{sec}}}\right) \\ &= \boxed{4.080 \text{ tons}} \end{aligned}$$

The answer is C.

3.3.

$$\text{COP} = \frac{\dot{Q}_L}{\dot{Q}_H - \dot{Q}_L}$$

$\dot{Q}_H$ is the rate of heat rejection to the surroundings (ambient air). Solve the preceding equation for $\dot{Q}_H$.

$$\begin{aligned} \dot{Q}_H &= \dot{Q}_L\left(\frac{\text{COP} + 1}{\text{COP}}\right) = \left(13.6 \frac{\text{Btu}}{\text{sec}}\right)\left(\frac{4+1}{4}\right) \\ &= \boxed{17.0 \text{ Btu/sec}} \end{aligned}$$

The answer is A.

3.4.

$$\begin{aligned} T_H &= 90°\text{F} + 460° = 550°\text{R} \\ T_L &= -40°\text{F} + 460° = 420°\text{R} \\ \text{COP} &= \frac{T_L}{T_H - T_L} = \frac{420°\text{R}}{550°\text{R} - 420°\text{R}} = 3.23 \\ \dot{Q}_L &= (\text{COP})\dot{W} \\ &= (3.23)\left(3.393 \frac{\text{Btu}}{\text{sec}}\right) \\ &= \left(10.96 \frac{\text{Btu}}{\text{sec}}\right)\left(\frac{1 \text{ ton}}{3.333 \frac{\text{Btu}}{\text{sec}}}\right) \\ &= \boxed{3.288 \text{ tons}} \end{aligned}$$

The answer is B.

SOLUTION 4

4.1. The change in internal energy is calculated as follows.

$$\begin{aligned}\Delta U &= mc_v\Delta T\\ &= (6\ \text{lbm})\left(0.361\ \frac{\text{Btu}}{\text{lbm-}^\circ\text{R}}\right)\\ &\quad\times\big((180^\circ\text{F}+460^\circ)-(70^\circ\text{F}+460^\circ)\big)\\ &= \boxed{238.3\ \text{Btu}}\end{aligned}$$

The answer is B.

4.2. The change in enthalpy is calculated as follows.

$$\begin{aligned}\Delta H &= mc_p\Delta T\\ &= (6\ \text{lbm})\left(0.427\ \frac{\text{Btu}}{\text{lbm-}^\circ\text{R}}\right)\\ &\quad\times\big((180^\circ\text{F}+460^\circ)-(70^\circ\text{F}+460^\circ)\big)\\ &= \boxed{281.8\ \text{Btu}}\end{aligned}$$

The answer is C.

4.3. The change in entropy is calculated using

$$\Delta S = mc_v\ln\left(\frac{T_2}{T_1}\right) + mR\ln\left(\frac{V_2}{V_1}\right)$$

For a constant volume process, $V_2 = V_1$.

Therefore,

$$\begin{aligned}\Delta S &= mc_v\ln\left(\frac{T_2}{T_1}\right)\\ T_2 &= 180^\circ\text{F}+460^\circ = 640^\circ\text{R}\\ T_1 &= 70^\circ\text{F}+460^\circ = 530^\circ\text{R}\\ \Delta S &= mc_v\ln\left(\frac{T_2}{T_1}\right)\\ &= (6\ \text{lbm})\left(0.361\ \frac{\text{Btu}}{\text{lbm-}^\circ\text{R}}\right)\ln\left(\frac{640^\circ\text{R}}{530^\circ\text{R}}\right)\\ &= \boxed{0.4084\ \text{Btu}/^\circ\text{R}}\end{aligned}$$

The answer is A.

4.4. For a constant-volume process, the work done, W, is

$$\int_{V_1}^{V_2} pdV = \boxed{0}$$

The answer is B.

4.5. The heat added is calculated by applying the first law for a closed system.

$$Q - W = \Delta U$$

Since $W = 0$,

$$Q = \Delta U = \boxed{238.3\ \text{Btu}}$$

The answer is A.

4.6. The enthalpy is calculated as follows.

$$H = U + pV$$

For a constant-pressure process,

$$\Delta H = \Delta U + p\Delta V$$

Applying the first law for a closed system, the heat added for a constant-pressure process is

$$\begin{aligned}Q &= \Delta U + W = \Delta U + p\Delta V\\ Q &= \Delta H\end{aligned}$$

From the solution to Prob. 4.2, $\Delta H = 281.8$ Btu.

Therefore, the heat added is $Q = \boxed{281.8\ \text{Btu.}}$

The answer is B.

4.7. The work done, W, is calculated by applying the first law for a closed system.

$$\begin{aligned}W &= Q - \Delta U\\ &= 281.8\ \text{Btu} - 238.3\ \text{Btu}\\ &= \boxed{43.5\ \text{Btu}}\end{aligned}$$

The answer is D.

4.8. The change in entropy, ΔS, is calculated as follows.

$$\Delta S = mc_p\ln\left(\frac{T_2}{T_1}\right) - mR\ln\left(\frac{p_2}{p_1}\right)$$

For a constant-pressure process, $p_2 = p_1$. Therefore,

$$\begin{aligned}\Delta S &= mc_p\ln\left(\frac{T_2}{T_1}\right)\\ &= (6\ \text{lbm})\left(0.427\ \frac{\text{Btu}}{\text{lbm-}^\circ\text{R}}\right)\ln\left(\frac{640^\circ\text{R}}{530^\circ\text{R}}\right)\\ &= \boxed{0.4832\ \text{Btu}/^\circ\text{R}}\end{aligned}$$

The answer is B.

4.9. Using the molar heat capacity data, the change in enthalpy, ΔH, is calculated as shown.

$$\Delta\overline{h} = \int_{T_1}^{T_2} \overline{c}_p dT$$

$$= \int_{530°\text{R}}^{640°\text{R}} \begin{pmatrix} 1.648 + 2.291 \times 10^{-2}\ T \\ - 0.4722 \times 10^{-5}\ T^2 \end{pmatrix} dT$$

$$= (1.648)(640°\text{R} - 530°\text{R}) + \left(\frac{2.291 \times 10^{-2}}{2}\right)\left((640°\text{R})^2 - (530°\text{R})^2\right) - \left(\frac{0.4722 \times 10^{-5}}{3}\right)\left((640°\text{R})^3 - (530°\text{R})^3\right)$$

$$= 1477\ \text{Btu/lbmol}$$

Note that $\overline{c}_p$ is in Btu/lbmol-°R and T is in °R.

$$\Delta h = \frac{\Delta\overline{h}}{M} = \frac{1477\ \frac{\text{Btu}}{\text{lbmol}}}{30\ \frac{\text{lbm}}{\text{lbmol}}} = 49.23\ \text{Btu/lbm}$$

$$\Delta H = m\Delta h = (6\ \text{lbm})\left(49.23\ \frac{\text{Btu}}{\text{lbm}}\right) = 295.4\ \text{Btu}$$

From the solution to Prob. 4.2, $\Delta H = 281.8$ Btu.

Therefore, the percent error is

$$\left(\frac{295.4\ \text{Btu} - 281.8\ \text{Btu}}{295.4\ \text{Btu}}\right) \times 100\% = \boxed{4.60\%}$$

The answer is C.

4.10. First Δh is determined for the temperature change from 70°F to 700°F as shown.

$$T_1 = 70°\text{F} + 460° = 530°\text{R}$$

$$T_1 = 700°\text{F} + 460° = 1160°\text{R}$$

$$\Delta\overline{h} = \int_{T_1}^{T_2} \overline{c}_p dT$$

$$= \int_{530°\text{R}}^{1160°\text{R}} \begin{pmatrix} 1.648 + 2.291 \times 10^{-2}\ T \\ - 0.4722 \times 10^{-5} T^2 \end{pmatrix} dT$$

$$= (1.648)(1160°\text{R} - 530°\text{R}) + \left(\frac{2.291 \times 10^{-2}}{2}\right)\left((1160°\text{R})^2 - (530°\text{R})^2\right) - \left(\frac{0.4722 \times 10^{-5}}{3}\right)\left((1160°\text{R})^3 - (530°\text{R})^3\right)$$

$$= 11{,}012\ \text{Btu/lbmol}$$

$$\Delta h = \frac{\Delta\overline{h}}{M} = \frac{11{,}012\ \frac{\text{Btu}}{\text{lbmol}}}{30\ \frac{\text{lbm}}{\text{lbmol}}} = 367.1\ \text{Btu/lbm}$$

$$\Delta h = c_{p_{\text{mean}}} \Delta T$$

$$c_{p_{\text{mean}}} = \frac{\Delta h}{\Delta T} = \frac{367.1\ \frac{\text{Btu}}{\text{lbm}}}{1160°\text{R} - 530°\text{R}} = \boxed{0.5827\ \text{Btu/lbm-°R}}$$

The answer is C.

Fluids

PROBLEM 1

Nitrogen at 140 psig and 550°F flows at a rate of 400 lbm/hr through 100 ft of 2 in, schedule-40 commercial steel pipe (ID = 2.067 in). The pipe includes a gate valve, four 90° elbows, three tees, and two check valves. The viscosity of nitrogen at flow conditions is 0.028 cP.

1.1. What is the flow velocity?

(A) 4 ft/sec
(B) 8 ft/sec
(C) 12 ft/sec
(D) 21 ft/sec

1.2. What is the Reynolds number?

(A) 22,000
(B) 33,000
(C) 38,000
(D) 44,000

1.3. What is the Darcy friction factor?

(A) 0.025
(B) 0.044
(C) 0.064
(D) 0.091

1.4. What is the head (friction) loss due to straight pipe?

(A) 1.1 in water
(B) 2.4 in water
(C) 6.8 in water
(D) 9.1 in water

1.5. What is the pressure drop due to fittings?

(A) 0.02 psi
(B) 0.04 psi
(C) 0.05 psi
(D) 0.09 psi

1.6. What is the total pressure drop in the system?

(A) 0.14 psi
(B) 0.31 psi
(C) 0.56 psi
(D) 0.91 psi

1.7. What is the contribution of fittings to the total pressure drop?

(A) 11%
(B) 16%
(C) 26%
(D) 37%

PROBLEM 2

A packed bed is used in a regeneration heater. The packing consists of cubes of 0.5 cm on each side. The height of the packing is 3.5 m. Air flows through the bed, entering at 40°C and 7 bars and leaving at 250°C. The mass velocity of the fluid is 3500 kg/h·m^2. The porosity of the bed is 0.44. The viscosity of air at flow conditions is 0.7×10^{-5} N·S/m^2.

2.1. What is the characteristic diameter of the particle?

(A) 0.002 m
(B) 0.003 m
(C) 0.004 m
(D) 0.005 m

2.2. What is the superficial velocity of air?

(A) 0.11 m/s
(B) 0.16 m/s
(C) 0.29 m/s
(D) 0.36 m/s

2.3. What is the Reynolds number for the particle?

(A) 310
(B) 520
(C) 690
(D) 830

2.4. At the stated conditions, the flow is

(A) laminar
(B) turbulent
(C) transient
(D) supersonic

2.5. What is the exit pressure of air?

(A) 6.01 bars
(B) 6.23 bars
(C) 6.39 bars
(D) 6.99 bars

PROBLEM 3

Hexene (SG = 0.68) is being pumped from a storage tank maintained at atmospheric pressure. The level of hexene in the tank varies from 40 ft to 6 ft above grade. The eye of the pump impeller is 2 ft above grade. Friction and other losses in the suction line amount to 1 lbf/in^2. The pump manufacturer has a specified NPSHR of 8 lbf/in^2. At operating conditions, the vapor pressure of hexene is 6 psia.

3.1. What is the NPSHR of the pump in feet of hexene?

(A) 18 ft
(B) 23 ft
(C) 27 ft
(D) 33 ft

3.2. What is the NPSHA of the pump in feet of hexene?

(A) 30 ft
(B) 34 ft
(C) 37 ft
(D) 39 ft

3.3. The values of NPSHR and NPSHA indicate which of the following?

(A) There will be vaporization of hexene during suction.
(B) There is sufficient suction head available.
(C) The level of hexene in the tank needs to be lowered.
(D) There will be backflow of hexene to the tank.

3.4. If the pump has an efficiency of 85% and is rated at 15 hp, what is the maximum head it can deliver while pumping 1200 gpm of hexene?

(A) 42 ft of hexene
(B) 49 ft of hexene
(C) 55 ft of hexene
(D) 62 ft of hexene

PROBLEM 4

A piping system is sketched as shown. All pipes are commercial steel, schedule-40. Water is flowing at 60°F. The total equivalent length for branch B is 1000 ft, and that for branch C is 1500 ft. Branch B is a 4 in pipe, and branch C is a 6 in pipe. f_C and f_B are Darcy friction factors. Q_A is 400 gpm. All pipes are at the same elevation.

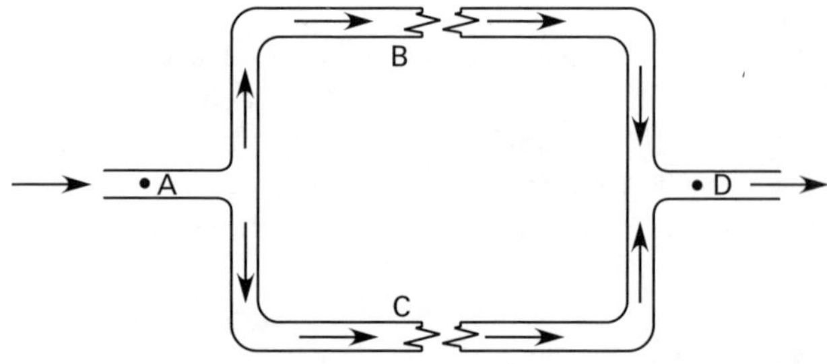

4.1. What is the velocity in branch C?

(A) 3.1 ft/sec
(B) 4.1 ft/sec
(C) 5.3 ft/sec
(D) 6.2 ft/sec

4.2. What is the volume flow rate in branch B?

(A) 100 gal/min
(B) 120 gal/min
(C) 140 gal/min
(D) 160 gal/min

4.3. If branch D is an 8 in pipe, what is the Darcy friction factor, f_D?

(A) 0.012
(B) 0.018
(C) 0.031
(D) 0.042

4.4. What is the pressure drop between A and D?

(A) 3.7 lbf/in^2
(B) 4.8 lbf/in^2
(C) 5.5 lbf/in^2
(D) 6.3 lbf/in^2

PROBLEM 5

Hydroelectric power is generated using water stored in a reservoir. The water level in the reservoir is maintained at an elevation of 4000 ft. The water intake is located 20 ft below the water level in the reservoir and leads to a 400 ft long, 6 in, schedule-40 galvanized iron pipe. This pipe exits to a turbine/generator unit (efficiency = 86%) located at an elevation of 3800 ft. The discharge pipe from the turbine is also a 6 in, schedule-40 galvanized iron pipe, 2100 ft in length, and it discharges into air at an elevation of 3700 ft. A flow rate of 450 gpm is maintained in the system using a control valve. The water temperature is 68°F. Minor losses amount to 700 pipe diameters.

5.1. What is the kinematic viscosity of water?

(A) 1.1×10^{-5} ft^2/sec
(B) 2.1×10^{-5} ft^2/sec
(C) 3.1×10^{-5} ft^2/sec
(D) 4.1×10^{-5} ft^2/sec

5.2. What is the velocity of flow in the pipe?

(A) 2.9 ft/sec
(B) 3.2 ft/sec
(C) 4.1 ft/sec
(D) 4.9 ft/sec

5.3. What is the Darcy friction factor?

(A) 0.011
(B) 0.022
(C) 0.032
(D) 0.043

5.4. What is the friction loss due to straight pipe?

(A) 27 ft
(B) 32 ft
(C) 40 ft
(D) 45 ft

5.5. What is the friction loss due to minor losses?

(A) 2.3 ft
(B) 3.1 ft
(C) 4.2 ft
(D) 5.7 ft

5.6. What is the total friction loss per pound of fluid?

(A) 38 ft-lbf/lbm
(B) 46 ft-lbf/lbm
(C) 51 ft-lbf/lbm
(D) 60 ft-lbf/lbm

5.7. What is the shaft work per pound of fluid?

(A) 220 ft-lbf/lbm
(B) 290 ft-lbf/lbm
(C) 330 ft-lbf/lbm
(D) 380 ft-lbf/lbm

5.8. What is the power generated?

(A) 13 kW
(B) 16 kW
(C) 18 kW
(D) 20 kW

Fluids Solutions

SOLUTION 1

1.1. First, calculate the density of nitrogen using the ideal gas law.

$$\rho = \frac{p}{RT}$$

$$p_{\text{gage}} = 140 \text{ lbf/in}^2$$

$$p_{\text{abs}} = p_{\text{gage}} + p_{\text{atm}} = 140 \frac{\text{lbf}}{\text{in}^2} + 14.7 \frac{\text{lbf}}{\text{in}^2} = 154.7 \text{ lbf/in}^2$$

R is the individual gas constant for $N_2 = \overline{R}/\text{MW}$. $\overline{R}$ is the universal gas constant, and MW is the molecular weight.

$$R = \frac{1545 \frac{\text{ft-lbf}}{\text{lbmol-°R}}}{28 \frac{\text{lbm}}{\text{lbmol}}} = 55.18 \text{ ft-lbf/lbm-°R}$$

$$T = 550°\text{F} + 460° = 1010°\text{R}$$

Therefore,

$$\rho = \frac{p}{RT} = \frac{\left(154.7 \frac{\text{lbf}}{\text{in}^2}\right)\left(144 \frac{\text{in}^2}{\text{ft}^2}\right)}{\left(55.18 \frac{\text{ft-lbf}}{\text{lbm-°R}}\right)(1010°\text{R})} = 0.3997 \text{ lbm/ft}^3$$

Next, calculate the volume flow rate of nitrogen.

$$\dot{V} = \frac{\dot{m}}{\rho} = \frac{\left(400 \frac{\text{lbm}}{\text{hr}}\right)\left(\frac{1 \text{ hr}}{60 \text{ min}}\right)\left(\frac{1 \text{ min}}{60 \text{ sec}}\right)}{0.3997 \frac{\text{lbm}}{\text{ft}^3}} = 0.2778 \text{ ft}^3/\text{sec}$$

Calculate the flow velocity.

The diameter is

$$D = \frac{2.067 \text{ in}}{12 \frac{\text{in}}{\text{ft}}} = 0.173 \text{ ft}$$

The velocity is

$$\text{v} = \frac{\dot{V}}{\frac{\pi D^2}{4}} = \frac{0.2778 \frac{\text{ft}^3}{\text{sec}}}{\left(\frac{\pi}{4}\right)(0.173 \text{ ft})^2} = \boxed{11.92 \text{ ft/sec}}$$

The answer is C.

1.2. The Reynolds number is

$$\text{Re} = \frac{D\text{v}\rho}{\mu}$$

The dynamic viscosity, μ, has units consistent with the units of D, v, and ρ.

$$\mu = (0.028 \text{ cP})\left(6.72 \times 10^{-4} \frac{\frac{\text{lbm}}{\text{ft-sec}}}{\text{cP}}\right) = 1.882 \times 10^{-5} \text{ lbm/ft-sec}$$

$$\text{Re} = \frac{D\text{v}\rho}{\mu} = \frac{(0.173 \text{ ft})\left(11.92 \frac{\text{ft}}{\text{sec}}\right)\left(0.3997 \frac{\text{lbm}}{\text{ft}^3}\right)}{1.882 \times 10^{-5} \frac{\text{lbm}}{\text{ft-sec}}} = \boxed{43{,}606}$$

The answer is D.

1.3. The roughness, ε, of steel pipe is 0.00015 ft.

$$\frac{\varepsilon}{D} = \frac{0.00015 \text{ ft}}{0.173 \text{ ft}} = 0.0009$$

From Moody's friction chart, the Darcy friction factor is $f = \boxed{0.0245.}$

The answer is A.

1.4. The head loss due to a straight pipe of 100 ft is calculated using Darcy's equation.

$$h_f = \frac{fLv^2}{2gD} = \frac{(0.0245)\,(100\text{ ft})\left(11.92\,\frac{\text{ft}}{\text{sec}}\right)^2}{(2)\left(32.2\,\frac{\text{ft}}{\text{sec}^2}\right)(0.173\text{ ft})}$$
$$= 31.25\text{ ft of N}_2$$

The friction loss in feet of water is

$$(h_f)_{H_2O} = \left(\frac{\rho_{N_2}}{\rho_{H_2O}}\right)(h_f)_{N_2} = \left(\frac{0.3997\,\frac{\text{lbm}}{\text{ft}^3}}{62.4\,\frac{\text{lbm}}{\text{ft}^3}}\right)(31.25\text{ ft})$$
$$= 0.2002\text{ ft of water}$$

The friction loss in inches of water is

$$(0.2002\text{ ft})\left(12\,\frac{\text{in}}{\text{ft}}\right) = \boxed{2.402\text{ in of water}}$$

The answer is B.

1.5. The loss coefficients for fittings are as follows.

fittings	quantity	K	$\sum K$
gate valve	1	0.20	0.20
90° elbow	4	0.70	2.80
tees (run)	3	0.4	1.2
check value	2	2.0	4.0
		total	8.20

The head loss due to the fittings is

$$h_m = \sum K\left(\frac{v^2}{2g}\right) = (8.20)\left(\frac{\left(11.92\,\frac{\text{ft}}{\text{sec}}\right)^2}{(2)\left(32.2\,\frac{\text{ft}}{\text{sec}^2}\right)}\right)$$
$$= 18.09\text{ ft of N}_2$$

The specific weight of nitrogen is

$$\gamma = \frac{\rho g}{g_c} = \frac{\left(0.3997\,\frac{\text{lbm}}{\text{ft}^3}\right)\left(32.2\,\frac{\text{ft}}{\text{sec}^2}\right)}{32.2\,\frac{\text{lbm-ft}}{\text{lbf-sec}^2}}$$
$$= 0.3997\text{ lbf/ft}^3$$

The pressure drop due to the fittings is

$$\Delta p_m = \gamma h_m = \left(0.3997\,\frac{\text{lbf}}{\text{ft}^3}\right)(18.09\text{ ft})\left(\frac{1\text{ ft}^2}{144\text{ in}^2}\right)$$
$$= \boxed{0.0502\text{ lbf/in}^2}$$

The answer is C.

1.6. The pressure drop due to the straight pipe is

$$\Delta p_f = \gamma h_f = \left(0.3997\,\frac{\text{lbf}}{\text{ft}^3}\right)(31.25\text{ ft})\left(\frac{1\text{ ft}^2}{144\text{ in}^2}\right)$$
$$= 0.0867\text{ lbf/in}^2$$

The total pressure drop in the system is the sum of the pressure drop in the pipe and fittings.

$$\Delta p = \Delta p_f + \Delta p_m = 0.0867\,\frac{\text{lbf}}{\text{in}^2} + 0.0502\,\frac{\text{lbf}}{\text{in}^2}$$
$$= \boxed{0.1369\text{ lbf/in}^2}$$

The answer is A.

1.7. The percentage of pressure drop due to fittings is

$$\left(\frac{\Delta p_m}{\Delta p}\right)\times 100\% = \left(\frac{0.0502\,\frac{\text{lbf}}{\text{in}^2}}{0.1369\,\frac{\text{lbf}}{\text{in}^2}}\right)\times 100\% = \boxed{36.67\%}$$

The answer is D.

SOLUTION 2

2.1. Since the particles are cubes, the characteristic diameter is determined by

$$D_p = \frac{6V_p}{A_p} = \left(\frac{(6)\,(0.5\text{ cm})^3}{(6)\,(0.5\text{ cm})^2}\right)\left(\frac{1\text{ m}}{100\text{ cm}}\right) = \boxed{0.005\text{ m}}$$

The answer is D.

2.2. Since the inlet and outlet temperatures vary, the density of air is determined at both the inlet and outlet conditions, and then the average of the two values is used in the calculations.

$$p = (7 \text{ bars})\left(\frac{10^5 \text{ Pa}}{1 \text{ bar}}\right) = 7\times10^5 \text{ Pa}$$

$$T_{\text{in}} = 40^\circ\text{C} + 273^\circ = 313\text{K}$$

$$\rho_{\text{in}} = \frac{p}{RT_{\text{in}}} = \frac{7\times10^5 \text{ Pa}}{\left(287 \frac{\text{J}}{\text{kg}\cdot\text{K}}\right)(313\text{K})}$$

$$= 7.792 \text{ kg/m}^3 \quad [\text{Note that J} = \text{N}\cdot\text{m}]$$

$$T_{\text{out}} = 250^\circ\text{C} + 273^\circ = 523\text{K}$$

$$\rho_{\text{out}} = \frac{p}{RT_{\text{out}}} = \frac{7\times10^5 \text{ Pa}}{\left(287 \frac{\text{J}}{\text{kg}\cdot\text{K}}\right)(523\text{K})}$$

$$= 4.664 \text{ kg/m}^3$$

The average density is

$$\rho = \frac{\rho_{\text{in}} + \rho_{\text{out}}}{2} = \frac{7.792 \frac{\text{kg}}{\text{m}^3} + 4.664 \frac{\text{kg}}{\text{m}^3}}{2} = 6.228 \text{ kg/m}^3$$

The superficial velocity is calculated using

$$\text{v}_s = \frac{G}{\rho}$$

G is mass flow rate per unit area (mass velocity), and ρ is density.

$$\text{v}_s = \frac{G}{\rho} = \left(\frac{3500 \frac{\text{kg}}{\text{h}\cdot\text{m}^2}}{6.228 \frac{\text{kg}}{\text{m}^3}}\right)\left(\frac{1 \text{ h}}{3600 \text{ s}}\right) = \boxed{0.1561 \text{ m/s}}$$

The answer is B.

2.3. The Reynolds number for the particle is

$$(\text{Re})_p = \frac{D_p \text{v}_s \rho}{\mu} = \frac{(0.005 \text{ m})\left(0.1561 \frac{\text{m}}{\text{s}}\right)\left(6.228 \frac{\text{kg}}{\text{m}^3}\right)}{0.7\times10^{-5} \frac{\text{N}\cdot\text{s}}{\text{m}^2}}$$

$$= \boxed{694.4} \quad [\text{Note that N} = \text{kg}\cdot\text{m/s}^2]$$

The answer is C.

2.4. $\frac{(\text{Re})_p}{1-\varepsilon} = \frac{694.4}{1-0.44} = 1240 > 5$

Therefore, the flow is $\boxed{\text{turbulent.}}$

The answer is B.

2.5. The pressure drop across the packed bed is calculated using the Ergun equation.

$$\Delta p = \left(\frac{1-\varepsilon}{\varepsilon^3}\right)\left(\frac{\rho \text{v}_s^2 L}{D_p}\right)\left(\frac{150(1-\varepsilon)}{(\text{Re})_p} + 1.75\right)$$

$$= \left(\frac{1-0.44}{(0.44)^3}\right) \times \left(\frac{\left(6.228 \frac{\text{kg}}{\text{m}^3}\right)\left(0.1561 \frac{\text{m}}{\text{s}}\right)^2(3.5 \text{ m})}{0.005 \text{ m}}\right) \times \left(\frac{(150)(1-0.44)}{694.4} + 1.75\right)\left(\frac{1 \text{ bar}}{10^5 \text{ Pa}}\right)$$

$$= 0.0131 \text{ bar}$$

$$p_{\text{exit}} = p_{\text{entrance}} - \Delta p = 7 \text{ bars} - 0.0131 \text{ bar}$$

$$= \boxed{6.9869 \text{ bars}}$$

The answer is D.

SOLUTION 3

3.1. The specific weight of hexene is

$$\gamma_{\text{hexene}} = (\text{SG})(\gamma_{\text{water}})$$

$$= (0.68)\left(62.4 \frac{\text{lbf}}{\text{ft}^3}\right)$$

$$= 42.43 \text{ lbf/ft}^3$$

The net positive suction head required is

$$\text{NPSHR} = \frac{p}{\gamma_{\text{hexene}}} = \frac{\left(8 \frac{\text{lbf}}{\text{in}^2}\right)\left(144 \frac{\text{in}^2}{\text{ft}^2}\right)}{42.43 \frac{\text{lbf}}{\text{ft}^3}}$$

$$= \boxed{27.15 \text{ ft of hexene}}$$

The answer is C.

3.2. Calculate the net positive suction head available using the minimum liquid level in the tank, that is, 6 ft. Since the eye of the impeller is 2 ft above grade, the liquid height in the suction line is 6 ft − 2 ft = 4 ft.

The pressure head at the source is

$$\begin{aligned} h &= \frac{p_{\text{source}}}{\gamma_{\text{hexene}}} \\ &= \frac{\left(14.7\ \frac{\text{lbf}}{\text{in}^2}\right)\left(144\ \frac{\text{in}^2}{\text{ft}^2}\right)}{42.43\ \frac{\text{lbf}}{\text{ft}^3}} \\ &= 49.88 \text{ ft of hexene} \end{aligned}$$

The head loss in the suction line is

$$\begin{aligned} h &= \frac{\Delta p}{\gamma_{\text{hexene}}} \\ &= \frac{\left(1\ \frac{\text{lbf}}{\text{in}^2}\right)\left(144\ \frac{\text{in}^2}{\text{ft}^2}\right)}{42.43\ \frac{\text{lbf}}{\text{ft}^3}} \\ &= 3.394 \text{ ft of hexene} \end{aligned}$$

The head due to vapor pressure is

$$\begin{aligned} h &= \frac{p_{\text{vapor}}}{\gamma_{\text{hexene}}} \\ &= \frac{\left(6\ \frac{\text{lbf}}{\text{in}^2}\right)\left(144\ \frac{\text{in}^2}{\text{ft}^2}\right)}{42.43\ \frac{\text{lbf}}{\text{ft}^3}} \\ &= 20.36 \text{ ft of hexene} \end{aligned}$$

$$\begin{aligned} \text{NPSHA} &= \text{source pressure head} + \text{liquid height} \\ &\quad - \text{head loss in suction line} \\ &\quad - \text{vapor pressure head} \\ &= 49.88 \text{ ft} + 4.000 \text{ ft} - 3.394 \text{ ft} - 20.36 \text{ ft} \\ &= \boxed{30.13 \text{ ft of hexene}} \end{aligned}$$

The answer is A.

3.3.

$$\begin{aligned} \text{NPSHA} - \text{NPSHR} &= 30.13 \text{ ft} - 27.15 \text{ ft} \\ &= 2.98 \text{ ft of hexene} \end{aligned}$$

There is sufficient suction head available.

The answer is B.

3.4. The horsepower delivered by the pump is

$$(\eta_{\text{pump}})(\text{rated hp}) = (0.85)(15 \text{ hp}) = 12.75 \text{ hp}$$

The volume flow rate, Q, is 1200 gpm.

$$\left(1200\ \frac{\text{gal}}{\text{min}}\right)\left(\frac{0.0022\ \frac{\text{ft}^3}{\text{sec}}}{1\ \frac{\text{gal}}{\text{min}}}\right) = 2.64 \text{ ft}^3/\text{sec}$$

The maximum head delivered by the pump in feet of hexene is

$$\begin{aligned} h &= \frac{\left(\frac{550\ \frac{\text{ft-lbf}}{\text{sec}}}{1 \text{ hp}}\right)(12.7 \text{ hp})}{Q\gamma} \\ &= \frac{\left(\frac{550\ \frac{\text{ft-lbf}}{\text{sec}}}{1 \text{ hp}}\right)(12.7 \text{ hp})}{\left(2.64\ \frac{\text{ft}^3}{\text{sec}}\right)\left(42.43\ \frac{\text{lbf}}{\text{ft}^3}\right)} \\ &= \boxed{62.36 \text{ ft of hexene}} \end{aligned}$$

The answer is D.

SOLUTION 4

4.1.

$$\begin{aligned} \Delta p = p_{\text{A}} - p_{\text{D}} &= \Delta p_{\text{B}} = \Delta p_{\text{C}} \\ \Delta p_{\text{B}} &= \gamma h_{f\text{B}} \\ \Delta p_{\text{C}} &= \gamma h_{f\text{C}} \\ h_{f\text{B}} &= h_{f\text{C}} \\ \frac{f_{\text{B}} L_{\text{B}} \text{v}_{\text{B}}^2}{2gD_{\text{B}}} &= \frac{f_{\text{C}} L_{\text{C}} \text{v}_{\text{C}}^2}{2gD_{\text{C}}} \\ \frac{\text{v}_{\text{B}}}{\text{v}_{\text{C}}} &= \sqrt{\left(\frac{f_{\text{C}}}{f_{\text{B}}}\right)\left(\frac{L_{\text{C}}}{L_{\text{B}}}\right)\left(\frac{D_{\text{B}}}{D_{\text{C}}}\right)} \end{aligned}$$

$f_{\text{B}} = f_{\text{C}}$ is a reasonable assumption for turbulent flow. Therefore,

$$\frac{\text{v}_{\text{B}}}{\text{v}_{\text{C}}} = \sqrt{\left(\frac{L_{\text{C}}}{L_{\text{B}}}\right)\left(\frac{D_{\text{B}}}{D_{\text{C}}}\right)}$$

From a table of pipe dimensions, the ID of a 4 in nominal pipe is $D_{\text{B}} = 0.3355$ ft and $A_{\text{B}} = 0.0884$ ft^2.

The ID of a 6 in nominal pipe is $D_C = 0.5054$ ft and $A_C = 0.2006\ \text{ft}^2$.

$$\begin{aligned}\frac{v_B}{v_C} &= \sqrt{\left(\frac{1500\ \text{ft}}{1000\ \text{ft}}\right)\left(\frac{0.3355\ \text{ft}}{0.5054\ \text{ft}}\right)}\\ &= 0.9979\\ v_B &= 0.9979 v_C\\ Q_A &= \left(400\ \frac{\text{gal}}{\text{min}}\right)\left(\frac{0.0022\ \frac{\text{ft}^3}{\text{sec}}}{1\ \frac{\text{gal}}{\text{min}}}\right)\\ &= 0.88\ \text{ft}^3/\text{sec}\\ Q_A &= Q_B + Q_C = v_B A_B + v_C A_C\\ &= 0.9979 v_C A_B + v_C A_C\\ &= v_C(0.9979 A_B + A_C)\end{aligned}$$

Therefore,

$$\begin{aligned}v_C &= \frac{Q_A}{0.9979 A_B + A_C}\\ &= \frac{0.88\ \frac{\text{ft}^3}{\text{sec}}}{(0.9979)\left(0.0884\ \text{ft}^2\right) + 0.2006\ \text{ft}^2}\\ &= \boxed{3.047\ \text{ft/sec}}\end{aligned}$$

The answer is A.

4.2.

$$\begin{aligned}v_B &= 0.9979 v_C = (0.9979)\left(3.047\ \frac{\text{ft}}{\text{sec}}\right)\\ &= 3.041\ \text{ft/sec}\\ Q_B &= v_B A_B\\ &= \left(3.041\ \frac{\text{ft}}{\text{sec}}\right)\left(0.0884\ \text{ft}^2\right)\\ &= \left(0.2688\ \frac{\text{ft}^3}{\text{sec}}\right)\left(\frac{1\ \frac{\text{gal}}{\text{min}}}{0.0022\ \frac{\text{ft}^3}{\text{sec}}}\right)\\ &= \boxed{122.2\ \text{gpm}}\end{aligned}$$

The answer is B.

4.3.

$$\begin{aligned}Q_D &= Q_A\\ &= 0.88\ \text{ft}^3/\text{sec}\\ v_D &= \frac{Q_D}{A_D}\end{aligned}$$

For an 8 in nominal pipe, $A_D = 0.3601\ \text{ft}^2$ and $D_D = 0.6771$ ft.

$$\begin{aligned}v_D &= \frac{0.88\ \frac{\text{ft}^3}{\text{sec}}}{0.3601\ \text{ft}^2} = 2.444\ \text{ft/sec}\\ (\text{Re})_D &= \frac{D_D v_D \rho}{\mu}\\ &= \frac{(0.6771\ \text{ft})\left(2.444\ \frac{\text{ft}}{\text{sec}}\right)\left(62.4\ \frac{\text{lbm}}{\text{ft}^3}\right)}{6.72\times10^{-4}\ \frac{\text{lbm}}{\text{ft-sec}}}\\ &= 1.54\times10^5\\ \frac{\varepsilon}{D_D} &= \frac{0.00015\ \text{ft}}{0.6771\ \text{ft}}\\ &= 0.0002\end{aligned}$$

From Moody's chart, $f_D = \boxed{0.0175.}$

The answer is B.

4.4.

$$\begin{aligned}(\text{Re})_B &= \frac{D_B v_B \rho}{\mu}\\ &= \frac{(0.3355\ \text{ft})\left(3.041\ \frac{\text{ft}}{\text{sec}}\right)\left(62.4\ \frac{\text{lbm}}{\text{ft}^3}\right)}{6.72\times10^{-4}\ \frac{\text{lbm}}{\text{ft-sec}}}\\ &= 9.47\times10^4\\ \frac{\varepsilon}{D_B} &= \frac{0.00015\ \text{ft}}{0.3355\ \text{ft}} = 0.0004\end{aligned}$$

From Moody's chart, $f_B = 0.0199$.

$$\begin{aligned}p_A - p_D &= \gamma h_{fB} = \gamma\left(\frac{f_B L_B v_B^2}{2g\ D_B}\right)\\ &= \left(62.4\ \frac{\text{lbf}}{\text{ft}^3}\right)\\ &\quad\times\left(\frac{(0.0199)(1000\ \text{ft})\left(3.041\ \frac{\text{ft}}{\text{sec}}\right)^2}{(2)\left(32.2\ \frac{\text{ft}}{\text{sec}^2}\right)(0.3355\ \text{ft})}\right)\\ &\quad\times\left(\frac{1\ \text{ft}^2}{144\ \text{in}^2}\right)\\ &= 3.691\ \text{psi}\end{aligned}$$

The answer is A.

SOLUTION 5

5.1. A sketch of the system is shown below.

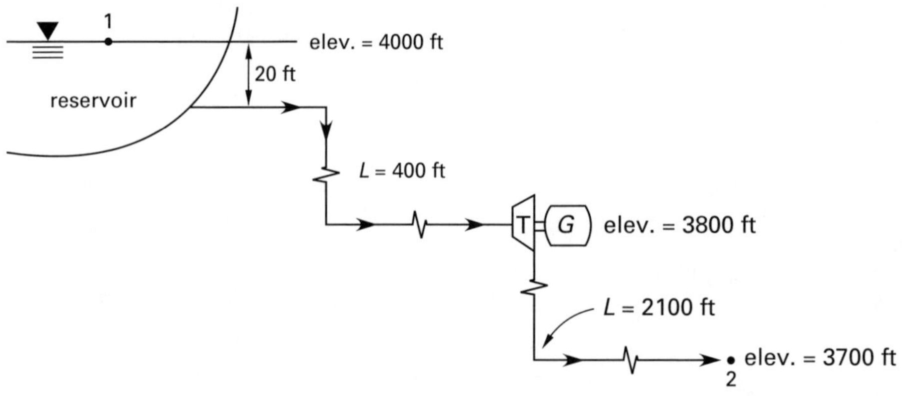

For water at 68°F,

$$\rho = 62.4\ \text{lbm/ft}^3$$

$$\gamma = 62.4\ \text{lbf/ft}^3$$

$$\mu = (1\ \text{cP})\left(\frac{6.72\times10^{-4}\ \frac{\text{lbm}}{\text{ft-sec}}}{1\ \text{cP}}\right) = 6.72\times10^{-4}\ \text{lbm/ft-sec}$$

The kinematic viscosity is

$$\upsilon = \frac{\mu}{\rho} = \frac{6.72\times10^{-4}\ \frac{\text{lbm}}{\text{ft-sec}}}{62.4\ \frac{\text{lbm}}{\text{ft}^3}} = \boxed{1.08\times10^{-5}\ \text{ft}^2/\text{sec}}$$

The answer is A.

5.2. The flow rate is

$$Q\left(450\ \frac{\text{gal}}{\text{min}}\right)\left(\frac{0.0022\ \frac{\text{ft}^3}{\text{sec}}}{1\ \frac{\text{gal}}{\text{min}}}\right) = 0.99\ \text{ft}^3/\text{sec}$$

The ID of a 6 in, schedule-40 pipe is

$$D = (6.065\ \text{in})\left(\frac{1\ \text{ft}}{12\ \text{in}}\right) = 0.5054\ \text{ft}$$

The flow velocity is

$$\text{v} = \frac{Q}{A} = \frac{Q}{\frac{\pi D^2}{4}} = \frac{0.99\ \frac{\text{ft}^3}{\text{sec}}}{\frac{\pi(0.5054\ \text{ft})^2}{4}} = \boxed{4.935\ \text{ft/sec}}$$

The answer is D.

5.3. The Reynolds number is

$$\text{Re} = \frac{D\text{v}\rho}{\mu} = \frac{(0.51\ \text{ft})\left(4.935\ \frac{\text{ft}}{\text{sec}}\right)\left(62.4\ \frac{\text{lbm}}{\text{ft}^3}\right)}{6.72\times10^{-4}\ \frac{\text{lbm}}{\text{ft-sec}}} = 2.337\times10^5$$

For galvanized iron pipe, the pipe roughness is $\varepsilon = 0.0005$ ft.

$$\frac{\varepsilon}{D} = \frac{0.0005\ \text{ft}}{0.5054\ \text{ft}} = 0.0010$$

From Moody's chart, the friction factor is $f = \boxed{0.0215.}$

The answer is B.

5.4. The head loss due to pipe friction is

$$h_f = \frac{fL\text{v}^2}{2gD} = \frac{(0.0215)(2500\ \text{ft})\left(4.935\ \frac{\text{ft}}{\text{sec}}\right)^2}{(2)\left(32.2\ \frac{\text{ft}}{\text{sec}^2}\right)(0.5054\ \text{ft})} = \boxed{40.22\ \text{ft}}$$

The answer is C.

5.5. The minor losses are calculated using the equivalent length of fittings.

$$L_e = 700D = (700)(0.5054\ \text{ft}) = 353.8\ \text{ft}$$

The minor losses are

$$h_m = \frac{fL_e\text{v}^2}{2gD} = \frac{(0.0215)(353.8\ \text{ft})\left(4.935\ \frac{\text{ft}}{\text{sec}}\right)^2}{(2)\left(32.2\ \frac{\text{ft}}{\text{sec}^2}\right)(0.5054\ \text{ft})}$$

$$= \boxed{5.692\ \text{ft}}$$

The answer is D.

5.6. The total friction loss is

$$h = h_f + h_m = 40.22\ \text{ft} + 5.692\ \text{ft} = 45.91\ \text{ft}$$

The friction loss per pound of fluid is

$$\frac{hg}{g_c} = \frac{(45.91\ \text{ft})\left(32.2\ \frac{\text{ft}}{\text{sec}^2}\right)}{32.2\ \frac{\text{lbm-ft}}{\text{lbf-sec}^2}}$$

$$= \boxed{45.91\ \text{ft-lbf/lbm}}$$

The answer is B.

5.7. The mechanical energy equation is applied between points 1 and 2 in the schematic.

$$\frac{p_1}{\gamma} + z_1 + \frac{\text{v}_1^2}{2g} + h_p = \frac{p_2}{\gamma} + z_2 + \frac{\text{v}_2^2}{2g} + h + h_t$$

$$p_1 = p_2 = 0$$

$$\text{v}_1 = \text{v}_2 = 0$$

$$h_p = 0$$

$$z_1 = 4000\ \text{ft}$$

$$z_2 = 3700\ \text{ft}$$

$$h = 45.91\ \text{ft}$$

Substitute the preceding values into the mechanical energy equation.

$$4000\ \text{ft} = 3700\ \text{ft} + 45.91\ \text{ft} + h_t$$

$$h_t = 254.1\ \text{ft}$$

The shaft work per pound of fluid is

$$\left(\frac{h_t g}{g_c}\right)\eta = \left(\frac{(254.1\ \text{ft})\left(32.2\ \frac{\text{ft}}{\text{sec}^2}\right)}{32.2\ \frac{\text{lbm-ft}}{\text{lbf-sec}^2}}\right)(0.86)$$

$$= \boxed{218.5\ \text{ft-lbf/lbm}}$$

The answer is A.

5.8. The power generated is

$$(Q\gamma h_t)\eta = \left(\left(0.99\ \frac{\text{ft}^3}{\text{sec}}\right)\left(62.4\ \frac{\text{lbf}}{\text{ft}^3}\right)(254.1\ \text{ft})(0.86)\right) \times \left(\frac{1\ \text{kW}}{737.6\ \frac{\text{ft-lbf}}{\text{sec}}}\right)$$

$$= \boxed{18.30\ \text{kW}}$$

The answer is C.

Heat Transfer

PROBLEM 1

A rotary furnace consists of a steel shell ($k = 25$ Btu/hr-ft-°F) with an inside diameter of 9 ft and a wall thickness of 2 in. The inside of the shell is lined with a 4.5 in layer of refractory brick ($k = 0.1$ Btu/hr-ft-°F). Inside the furnace are hot gases at an average temperature of 1800°F. The inside heat transfer coefficient is 19 Btu/hr-ft^2-°F. The outside of the steel shell is lined with a 3.5 in thick rock wool insulation ($k = 0.025$ Btu/hr-ft-°F). The ambient temperature is 70°F, and the furnace is 45 ft long. The outside heat transfer coefficient is 2.28 Btu/hr-ft^2-°F. Rectangular geometry can be assumed because of the large diameter and small thickness of each layer. The outside area of the furnace with insulation can be used as the basis for all calculations.

1.1. What is the heat transfer area?

(A) 1200 ft^2
(B) 1300 ft^2
(C) 1400 ft^2
(D) 1500 ft^2

1.2. The greatest resistance to heat transfer is due to the

(A) steel shell
(B) ambient air
(C) refractory brick
(D) rock wool insulation

1.3. What is the overall resistance to heat transfer?

(A) 0.006 hr-°F/Btu
(B) 0.011 hr-°F/Btu
(C) 0.027 hr-°F/Btu
(D) 0.051 hr-°F/Btu

1.4. What is the heat loss from the furnace?

(A) 1.5×10^5 Btu/hr
(B) 2.2×10^5 Btu/hr
(C) 2.7×10^5 Btu/hr
(D) 3.9×10^5 Btu/hr

1.5. If the safe operating temperature of the refractory brick is 1900°F, what is the margin of safety?

(A) 106°F
(B) 145°F
(C) 193°F
(D) 204°F

1.6. What is the highest temperature attained by the insulation?

(A) 1120°F
(B) 1380°F
(C) 1670°F
(D) 1810°F

1.7. What is the overall heat transfer coefficient?

(A) 0.032 Btu/hr-ft^2-°F
(B) 0.041 Btu/hr-ft^2-°F
(C) 0.051 Btu/hr-ft^2-°F
(D) 0.063 Btu/hr-ft^2-°F

1.8. What is the additional insulation thickness required to reduce the heat loss by 25%?

(A) 1.1 in
(B) 1.6 in
(C) 2.2 in
(D) 2.7 in

PROBLEM 2

Water at 195°F and at a mass flow rate of 5000 lbm/hr is used in heating ethylene glycol that enters a heat exchanger at 65°F. A process downstream requires 12,000 lbm/hr of ethylene glycol at 100°F. Use the following data.

outer pipe:

$$\text{ID} = 0.1674 \text{ ft}$$

inner pipe:

$$\text{ID} = 0.1076 \text{ ft}$$
$$\text{OD} = 0.1146 \text{ ft}$$

properties of water:

$$\text{SG} = 0.98$$
$$c_p = 0.9994 \text{ Btu/lbm-°F}$$
$$\upsilon = 0.514 \times 10^{-5} \text{ ft}^2/\text{sec}$$
$$k = 0.376 \text{ Btu/hr-ft-°F}$$

properties of ethylene glycol:

$$\text{SG} = 1.1$$
$$c_p = 0.612 \text{ Btu/lbm-°F}$$
$$\upsilon = 5.11 \times 10^{-5} \text{ ft}^2/\text{sec}$$
$$k = 0.150 \text{ Btu/hr-ft-°F}$$

For Probs. 2.1 through 2.6, assume a counterflow, double-pipe heat exchanger.

2.1. At the stated conditions,

(A) water should be routed through the inner pipe
(B) glycol should be routed through the inner pipe
(C) either water or glycol could be routed through the inner pipe
(D) water should be routed through the outer pipe

2.2. What is the heat exchanger load?

(A) 196,000 Btu/hr
(B) 220,000 Btu/hr
(C) 231,000 Btu/hr
(D) 257,000 Btu/hr

2.3. What is the log mean temperature difference (LMTD)?

(A) 68°F
(B) 76°F
(C) 87°F
(D) 96°F

2.4. What is the inside heat transfer coefficient?

(A) 520 Btu/hr-ft^2-°F
(B) 670 Btu/hr-ft^2-°F
(C) 780 Btu/hr-ft^2-°F
(D) 890 Btu/hr-ft^2-°F

2.5. What is the outside heat transfer coefficient?

(A) 110 Btu/hr-ft^2-°F
(B) 170 Btu/hr-ft^2-°F
(C) 260 Btu/hr-ft^2-°F
(D) 310 Btu/hr-ft^2-°F

2.6. What is the length of the exchanger?

(A) 46 ft
(B) 51 ft
(C) 62 ft
(D) 75 ft

2.7. If parallel flow is used instead of counterflow, what is the length of the exchanger?

(A) 51 ft
(B) 65 ft
(C) 75 ft
(D) 82 ft

For Probs. 2.8 and 2.9, consider a 1-2 shell-and-tube heat exchanger (one shell pass and two tube passes) with an overall heat transfer coefficient of 95 Btu/hr-ft^2-°F and counterflow with water on the tube side and ethylene glycol on the shell side.

2.8. What is the mean temperature difference (MTD) correction factor?

(A) 0.81
(B) 0.87
(C) 0.92
(D) 0.97

2.9. What is the heat exchanger area?

(A) 25 ft^2
(B) 28 ft^2
(C) 32 ft^2
(D) 35 ft^2

For Probs. 2.10–2.13, consider the double-pipe heat exchanger with counterflow and the same area as in Prob. 2.6. The glycol entrance temperature is lowered to 50°F. The water entrance temperature remains unchanged, and the mass flow rates of water and glycol remain the same.

2.10. What is the heat exchanger effectiveness?

(A) 29%
(B) 39%
(C) 48%
(D) 54%

2.11. What is the heat exchanger load?

(A) 208,000 Btu/hr
(B) 238,000 Btu/hr
(C) 257,000 Btu/hr
(D) 286,000 Btu/hr

2.12. What is the exit temperature of the ethylene glycol?

(A) 69°F
(B) 77°F
(C) 89°F
(D) 97°F

2.13. What is the exit temperature of the water?

(A) 130°F
(B) 140°F
(C) 150°F
(D) 160°F

PROBLEM 3

A thermocouple is used to measure the temperature of a gas flowing through a pipe. The pipe wall is at 700°F, and the thermocouple indicates a temperature of 300°F. The heat transfer coefficient from the thermocouple surface to the gas is 40 Btu/hr-ft^2-°F. The emissivity of the thermocouple material is 0.5. The actual temperature of the gas is

(A) 270°F
(B) 310°F
(C) 330°F
(D) 350°F

Heat Transfer Solutions

SOLUTION 1

1.1. The outside diameter of the furnace is the heat transfer area. The thickness of each layer represents an increase in radius and must be multiplied by 2 to obtain the diameter. Therefore, the outside diameter is

$$D_o = 9 \text{ ft} + (2)\left(\frac{2 \text{ in}}{12 \frac{\text{in}}{\text{ft}}}\right) + (2)\left(\frac{3.5 \text{ in}}{12 \frac{\text{in}}{\text{ft}}}\right) = 9.917 \text{ ft}$$

$$A_o = \pi D_o L = \pi(9.917 \text{ ft})(45 \text{ ft}) = \boxed{1402 \text{ ft}^2}$$

The answer is C.

1.2. Using rectangular geometry, sketch the system and the temperature profile as shown.

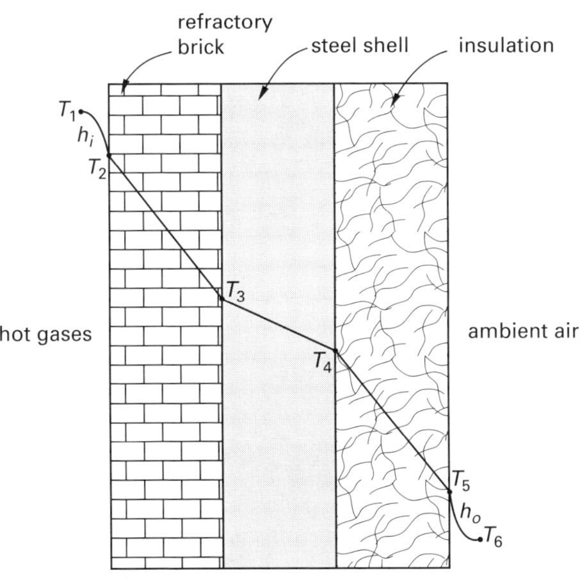

The heat transfer resistances per unit area are calculated as shown.

The convection resistance due to hot gases is

$$\frac{1}{h_i} = \frac{1}{19 \frac{\text{Btu}}{\text{hr-ft}^2\text{-}^\circ\text{F}}} = 0.0526 \text{ hr-ft}^2\text{-}^\circ\text{F/Btu}$$

The thickness of the steel shell is

$$(\Delta x)_s = \frac{2 \text{ in}}{12 \frac{\text{in}}{\text{ft}}} = 0.1667 \text{ ft}$$

The conduction resistance of the steel shell is

$$\left(\frac{\Delta x}{k}\right)_s = \frac{0.1667 \text{ ft}}{25 \frac{\text{Btu}}{\text{hr-ft-}^\circ\text{F}}} = 0.0067 \text{ hr-ft-}^\circ\text{F/Btu}$$

The thickness of the refractory brick is

$$(\Delta x)_b = \frac{4.5 \text{ in}}{12 \frac{\text{in}}{\text{ft}}} = 0.3750 \text{ ft}$$

The conduction resistance of the refractory brick is

$$\left(\frac{\Delta x}{k}\right)_b = \frac{0.3750 \text{ ft}}{0.1 \frac{\text{Btu}}{\text{hr-ft-}^\circ\text{F}}} = 3.750 \text{ hr-ft-}^\circ\text{F/Btu}$$

The thickness of the insulating layer is

$$(\Delta x)_{\text{ins}} = \frac{3.5 \text{ in}}{12 \frac{\text{in}}{\text{ft}}} = 0.2917 \text{ ft}$$

The conduction resistance of the insulating layer is

$$\left(\frac{\Delta x}{k}\right)_{\text{ins}} = \frac{0.2917 \text{ ft}}{0.025 \frac{\text{Btu}}{\text{hr-ft-}^\circ\text{F}}} = 11.67 \text{ hr-ft-}^\circ\text{F/Btu}$$

The convection resistance due to ambient air is

$$\frac{1}{h_o} = \frac{1}{2.28 \frac{\text{Btu}}{\text{hr-ft}^2\text{-}^\circ\text{F}}} = 0.4386 \text{ hr-ft}^2\text{-}^\circ\text{F/Btu}$$

> The highest resistance to heat transfer is due to the insulating layer.

The answer is D.

1.3. The overall resistance to heat transfer is

$$\Sigma R = \left(\frac{1}{A}\right)\left(\frac{1}{h_i} + \left(\frac{\Delta x}{k}\right)_b + \left(\frac{\Delta x}{k}\right)_s + \left(\frac{\Delta x}{k}\right)_{\text{ins}} + \frac{1}{h_o}\right)$$

$$= \left(\frac{1}{1402 \text{ ft}^2}\right)\left(\begin{array}{l} 0.0526 \, \frac{\text{hr-ft}^2\text{-}^\circ\text{F}}{\text{Btu}} \\ + 3.750 \, \frac{\text{hr-ft}^2\text{-}^\circ\text{F}}{\text{Btu}} \\ + 0.0067 \, \frac{\text{hr-ft}^2\text{-}^\circ\text{F}}{\text{Btu}} \\ + 11.67 \, \frac{\text{hr-ft}^2\text{-}^\circ\text{F}}{\text{Btu}} \\ + 0.4386 \, \frac{\text{hr-ft}^2\text{-}^\circ\text{F}}{\text{Btu}} \end{array}\right)$$

$$= \boxed{0.0114 \text{ hr-}^\circ\text{F/Btu}}$$

The answer is B.

1.4. $$q = \frac{\Delta T_{\text{overall}}}{\Sigma R} = \frac{T_1 - T_6}{\Sigma R} = \frac{1800^\circ\text{F} - 70^\circ\text{F}}{0.0114 \, \frac{\text{hr-}^\circ\text{F}}{\text{Btu}}}$$

$$= \boxed{1.518 \times 10^5 \text{ Btu/hr}}$$

The answer is A.

1.5. The temperature at the brick surface, T_2, can be calculated as follows.

$$q = \frac{T_1 - T_2}{\frac{1}{h_i A}}$$

$$T_1 - T_2 = q\left(\frac{1}{h_i A}\right)$$

$$= \left(1.518 \times 10^5 \, \frac{\text{Btu}}{\text{hr}}\right) \times \left(\frac{1}{\left(19 \, \frac{\text{Btu}}{\text{hr-ft}^2\text{-}^\circ\text{F}}\right)(1402 \text{ ft}^2)}\right)$$

$$= 6^\circ\text{F}$$

$$T_2 = T_1 - 6^\circ\text{F} = 1800^\circ\text{F} - 6^\circ\text{F} = 1794^\circ\text{F}$$

The margin of safety is

$$1900^\circ\text{F} - 1794^\circ\text{F} = \boxed{106^\circ\text{F}}$$

The answer is A.

1.6. The highest temperature attained by the insulation, T_4, is calculated as shown.

$$q = \frac{T_5 - T_6}{\frac{1}{h_o A}}$$

$$T_5 - T_6 = q\left(\frac{1}{h_o A}\right)$$

$$= \left(1.518 \times 10^5 \, \frac{\text{Btu}}{\text{hr}}\right) \times \left(\frac{1}{\left(2.28 \, \frac{\text{Btu}}{\text{hr-ft}^2\text{-}^\circ\text{F}}\right)(1402 \text{ ft}^2)}\right)$$

$$= 47^\circ\text{F}$$

$$T_5 - 70^\circ\text{F} = 47^\circ\text{F}$$

$$T_5 = 117^\circ\text{F}$$

$$q = \frac{T_4 - T_5}{\left(\frac{\Delta x}{kA}\right)_{\text{ins}}}$$

$$T_4 - T_5 = q\left(\frac{\Delta x}{k}\right)_{\text{ins}}\left(\frac{1}{A}\right)$$

$$= \left(1.518 \times 10^5 \, \frac{\text{Btu}}{\text{hr}}\right)\left(11.67 \, \frac{\text{hr-ft}^2\text{-}^\circ\text{F}}{\text{Btu}}\right) \times \left(\frac{1}{1402 \text{ ft}^2}\right)$$

$$= 1264^\circ\text{F}$$

$$T_4 = 1264^\circ\text{F} + T_5 = 1264^\circ\text{F} + 117^\circ\text{F}$$

$$= \boxed{1381^\circ\text{F}}$$

The answer is B.

1.7. $q = UA\Delta T_{\text{overall}}$

$$U = \frac{q}{A\Delta T_{\text{overall}}}$$

$$= \frac{1.518 \times 10^5 \, \frac{\text{Btu}}{\text{hr}}}{(1402 \text{ ft}^2)(1800^\circ\text{F} - 70^\circ\text{F})}$$

$$= \boxed{0.0626 \text{ Btu/hr-ft}^2\text{-}^\circ\text{F}}$$

The answer is D.

1.8. The new value for the heat loss is

$$q_{\text{new}} = 0.75 q_{\text{old}} = (0.75)\left(1.518 \times 10^5 \, \frac{\text{Btu}}{\text{hr}}\right)$$

$$= 1.138 \times 10^5 \text{ Btu/hr}$$

If t is the additional thickness of insulation needed, then

$$q_{\text{new}} = \frac{\Delta T_{\text{overall}}}{(\Sigma R)_{\text{old}} + \left(\frac{t}{kA}\right)_{\text{ins}}}$$

$$\begin{aligned}\left(\frac{t}{kA}\right)_{\text{ins}} &= \frac{\Delta T_{\text{overall}}}{q_{\text{new}}} - \Sigma R_{\text{old}} \\ &= \frac{1800^\circ\text{F} - 70^\circ\text{F}}{1.138 \times 10^5 \ \frac{\text{Btu}}{\text{hr}}} - 0.0114 \ \frac{\text{hr-}^\circ\text{F}}{\text{Btu}} \\ &= 0.0038 \ \text{hr-}^\circ\text{F/Btu}\end{aligned}$$

$$\begin{aligned}t &= \left(0.0038 \ \frac{\text{hr-}^\circ\text{F}}{\text{Btu}}\right)\left(0.025 \ \frac{\text{Btu}}{\text{hr-ft-}^\circ\text{F}}\right) \\ &\quad \times \left(1402 \ \text{ft}^2\right)\left(12 \ \frac{\text{in}}{\text{ft}}\right) \\ &= \boxed{1.6 \ \text{in}}\end{aligned}$$

The answer is B.

SOLUTION 2

2.1. The fluid with the higher flow rate should be routed through the section with the larger area. Therefore, the cross-sectional areas of the inner pipe and the annular region are calculated first.

$$\begin{aligned}D_{ii} &= \text{ID of inner pipe} \\ D_{io} &= \text{OD of inner pipe} \\ D_{oi} &= \text{ID of outer pipe}\end{aligned}$$

The cross-sectional area of the inner pipe is

$$A_p = \frac{\pi}{4}D_{ii}^2 = \left(\frac{\pi}{4}\right)(0.1076 \ \text{ft})^2 = 0.0091 \ \text{ft}^2$$

The cross-sectional area of the annular region is

$$\begin{aligned}A_a &= \left(\frac{\pi}{4}\right)\left(D_{oi}^2 - D_{io}^2\right) \\ &= \left(\frac{\pi}{4}\right)\left((0.1674 \ \text{ft})^2 - (0.1146 \ \text{ft})^2\right) \\ &= 0.0117 \ \text{ft}^2\end{aligned}$$

Therefore, ethylene glycol is routed through the annular region, and water is routed through the inner pipe.

The answer is A.

2.2. Glycol is the cold fluid, hence the subscript c is used.

The heat gained by glycol is equal to the heat exchanger load, q.

$$\begin{aligned}q &= \dot{m}_c c_{pc} \Delta T_c \\ &= \left(12{,}000 \ \frac{\text{lbm}}{\text{hr}}\right)\left(0.612 \ \frac{\text{Btu}}{\text{lbm-}^\circ\text{F}}\right)(100^\circ\text{F} - 65^\circ\text{F}) \\ &= \boxed{257{,}040 \ \text{Btu/hr}}\end{aligned}$$

The answer is D.

2.3. The heat balance equation is as follows.

$$\begin{aligned}\text{heat gained by glycol} &= \text{heat lost by water} \\ &= q\text{, the heat exchanger load}\end{aligned}$$

Since water is the hot fluid, the subscript h is used.

$$q = 257{,}040 \ \frac{\text{Btu}}{\text{hr}} = \dot{m}_h c_{ph} \Delta T_h$$

$$\begin{aligned}\Delta T_h &= \frac{257{,}040 \ \frac{\text{Btu}}{\text{hr}}}{\dot{m}_h c_{ph}} \\ &= \frac{257{,}040 \ \frac{\text{Btu}}{\text{hr}}}{\left(5000 \ \frac{\text{lbm}}{\text{hr}}\right)\left(0.9994 \ \frac{\text{Btu}}{\text{lbm-}^\circ\text{F}}\right)} = 51^\circ\text{F}\end{aligned}$$

The inlet temperature of the water is 195°F. Therefore, the exit temperature of water is

$$T_{\text{inlet,water}} - \Delta T_h = 195^\circ\text{F} - 51^\circ\text{F} = 144^\circ\text{F}$$

The flow schematic for counter-flow is shown.

195°F 1 water 2 144°F

100°F ethylene glycol 65°F

ΔT_1 = 95°F ΔT_2 = 79°F

$$\Delta T_{\text{LMTD}} = \frac{\Delta T_1 - \Delta T_2}{\ln\left(\frac{\Delta T_1}{\Delta T_2}\right)} = \frac{95^\circ\text{F} - 79^\circ\text{F}}{\ln\left(\frac{95^\circ\text{F}}{79^\circ\text{F}}\right)} = \boxed{86.75^\circ\text{F}}$$

The answer is C.

2.4. Subscripts p, i, and w, denoting pipe, inside, and water, respectively, are used for the inner pipe. The inside heat transfer coefficient is determined as follows.

$$\begin{aligned}
\text{v}_w &= \frac{\dot{m}_w}{\rho_w A_p} \\
&= \frac{\left(5000\ \frac{\text{lbm}}{\text{hr}}\right)\left(\frac{1\ \text{hr}}{3600\ \text{sec}}\right)}{(0.98)\left(62.4\ \frac{\text{lbm}}{\text{ft}^3}\right)(0.0091\ \text{ft}^2)} \\
&= 2.496\ \text{ft/sec}
\end{aligned}$$

$$\begin{aligned}
(\text{Re})_w &= \frac{D_p \text{v}_w}{\upsilon_w} \\
&= \frac{(0.1076\ \text{ft})\left(2.496\ \frac{\text{ft}}{\text{sec}}\right)}{0.514\times 10^{-5}\ \frac{\text{ft}^2}{\text{sec}}} \\
&= 52{,}251
\end{aligned}$$

$$\begin{aligned}
(\text{Pr})_w &= \frac{c_{pw}\mu_w}{k_w} = \frac{c_{pw}\rho_w \upsilon_w}{k_w} \\
&= \frac{\left(0.9994\ \frac{\text{Btu}}{\text{lbm-}^\circ\text{F}}\right)(0.98)\left(62.4\ \frac{\text{lbm}}{\text{ft}^3}\right) \times \left(0.514\times 10^{-5}\ \frac{\text{ft}^2}{\text{sec}}\right)}{\left(0.376\ \frac{\text{Btu}}{\text{hr-ft-}^\circ\text{F}}\right)\left(\frac{1\ \text{hr}}{3600\ \text{sec}}\right)} \\
&= 3.00
\end{aligned}$$

Water is being cooled. Therefore,

$$\begin{aligned}
(\text{Nu})_w &= (0.023)(\text{Re})_w^{0.8}(\text{Pr})_w^{0.3} \\
&= (0.023)\left(52{,}251\right)^{0.8}\left(3.00\right)^{0.3} \\
&= 190.3
\end{aligned}$$

$$(\text{Nu})_w = \frac{h_i D_p}{k_w} = 190.3$$

$$\begin{aligned}
h_i &= \frac{190.3 k_w}{D_p} = \frac{(190.3)\left(0.376\ \frac{\text{Btu}}{\text{hr-ft-}^\circ\text{F}}\right)}{0.1076\ \text{ft}} \\
&= \boxed{665.0\ \text{Btu/hr-ft}^2\text{-}^\circ\text{F}}
\end{aligned}$$

The answer is B.

2.5. Subscripts a, o, and g, denoting annulus, outside, and glycol, respectively, are used for the annular region. The outside heat transfer coefficient is determined as follows.

$$\begin{aligned}
\text{v}_g &= \frac{\dot{m}_g}{\rho_g A_a} \\
&= \frac{\left(12{,}000\ \frac{\text{lbm}}{\text{hr}}\right)\left(\frac{1\ \text{hr}}{3600\ \text{sec}}\right)}{(1.1)\left(62.4\ \frac{\text{lbm}}{\text{ft}^3}\right)(0.0117\ \text{ft}^2)} \\
&= 4.151\ \text{ft/sec}
\end{aligned}$$

Since this is an annular region, the equivalent diameter must be calculated.

$$\begin{aligned}
D_e &= 4R_H = (4)\left(\frac{\text{cross-sectional area}}{\text{wetted perimeter}}\right) \\
&= (4)\left(\frac{0.0117\ \text{ft}^2}{\pi\,(0.1146\ \text{ft} + 0.1674\ \text{ft})}\right) \\
&= 0.0528\ \text{ft}
\end{aligned}$$

$$(\text{Re})_g = \frac{D_e \text{v}_g}{\upsilon_g} = \frac{(0.0528\ \text{ft})\left(4.151\ \frac{\text{ft}}{\text{sec}}\right)}{5.11\times 10^{-5}\ \frac{\text{ft}^2}{\text{sec}}} = 4289$$

$$\begin{aligned}
(\text{Pr})_g &= \frac{c_{pg}\mu_g}{k_g} = \frac{c_{pg}\rho_g \upsilon_g}{k_g} \\
&= \frac{\left(0.612\ \frac{\text{Btu}}{\text{lbm-}^\circ\text{F}}\right)(1.1)\left(62.4\ \frac{\text{lbm}}{\text{ft}^3}\right) \times \left(5.11\times 10^{-5}\ \frac{\text{ft}^2}{\text{sec}}\right)}{\left(0.150\ \frac{\text{Btu}}{\text{hr-ft-}^\circ\text{F}}\right)\left(\frac{1\ \text{hr}}{3600\ \text{sec}}\right)} \\
&= 52
\end{aligned}$$

Glycol is being heated. Therefore,

$$\begin{aligned}
(\text{Nu})_g &= (0.023)(\text{Re})_g^{0.8}(\text{Pr})_g^{0.4} \\
&= (0.023)\left(4289\right)^{0.8}\left(52\right)^{0.4} \\
&= 89.95
\end{aligned}$$

$$(\text{Nu})_g = \frac{h_o D_e}{k_g} = 89.95$$

$$\begin{aligned}
h_o &= \frac{89.95 k_g}{D_e} = \frac{(89.95)\left(0.150\ \frac{\text{Btu}}{\text{hr-ft-}^\circ\text{F}}\right)}{0.0528\ \text{ft}} \\
&= \boxed{255.5\ \text{Btu/hr-ft}^2\text{-}^\circ\text{F}}
\end{aligned}$$

The answer is C.

2.6. The overall heat transfer coefficient is determined using the following equation.

$$\frac{1}{U_i} = \frac{1}{h_i} + \frac{A_i \ln\left(\frac{r_o}{r_i}\right)}{2\pi k_p L} + \left(\frac{A_i}{A_o}\right)\left(\frac{1}{h_o}\right) \qquad \text{[I]}$$

The thermal conductivity of pipe material (steel) is $k_p = 25$ Btu/hr-ft-°F. All dimensions are those of the inner pipe.

$$A_i = \pi D_i L = \pi (0.1076) L = 0.3380L$$

$$\frac{A_i}{A_o} = \frac{\pi D_i L}{\pi D_o L} = \frac{D_i}{D_o} = \frac{0.1076 \text{ ft}}{0.1146 \text{ ft}} = 0.9389$$

$$\frac{r_o}{r_i} = \frac{D_o}{D_i} = \frac{1}{0.9389} = 1.065$$

Substituting into Eq. I,

$$\frac{1}{U_i} = \frac{1}{665 \frac{\text{Btu}}{\text{hr-ft}^2\text{-°F}}} + \frac{(0.3380L)\ln(1.065)}{2\pi\left(25 \frac{\text{Btu}}{\text{hr-ft-°F}}\right)L} + \frac{0.9389}{255.5 \frac{\text{Btu}}{\text{hr-ft}^2\text{-°F}}}$$

$$= 0.0053 \text{ hr-ft}^2\text{-°F/Btu}$$

$$U_i = \frac{1}{0.0053 \frac{\text{hr-ft}^2\text{-°F}}{\text{Btu}}} = 188.7 \text{ Btu/hr-ft}^2\text{-°F}$$

$$q = U_i A_i \Delta T_{\text{LMTD}}$$

$$A_i = \frac{q}{U_i \Delta T_{\text{LMTD}}} = \frac{257{,}040 \frac{\text{Btu}}{\text{hr}}}{\left(188.7 \frac{\text{Btu}}{\text{hr-ft}^2\text{-°F}}\right)(87\text{°F})}$$

$$= 15.66 \text{ ft}^2$$

$$A_i = \pi D_i L$$

$$L = \frac{A_i}{\pi D_i} = \frac{15.66 \text{ ft}^2}{\pi (0.1076 \text{ ft})} = \boxed{46.33 \text{ ft}}$$

The answer is A.

2.7. The flow schematic for parallel flow is as follows.

195°F	1 —— water ——→ 2	144°F
65°F	—— ethylene glycol ——→	100°F
ΔT_1 = 130°F		ΔT_2 = 44°F

$$\Delta T_{\text{LMTD}} = \frac{\Delta T_1 - \Delta T_2}{\ln\left(\frac{\Delta T_1}{\Delta T_2}\right)} = \frac{130\text{°F} - 44\text{°F}}{\ln\left(\frac{130\text{°F}}{44\text{°F}}\right)} = 79.38\text{°F}$$

$$A_i = \frac{q}{U_i \Delta T_{\text{LMTD}}} = \frac{257{,}040 \frac{\text{Btu}}{\text{hr}}}{\left(188.7 \frac{\text{Btu}}{\text{hr-ft}^2\text{-°F}}\right)(79.38\text{°F})} = 17.16 \text{ ft}^2$$

$$L = \frac{A_i}{\pi D_i} = \frac{17.16 \text{ ft}^2}{\pi (0.1076 \text{ ft})} = \boxed{50.76 \text{ ft}}$$

The answer is A.

2.8. From the TEMA MTD correction factor chart for a 1-2 exchanger, the following information is obtained.

T_1 = shell side entrance temperature = 65°F (glycol)
T_2 = shell side exit temperature = 100°F (glycol)
t_1 = tube side entrance temperature = 195°F (water)
t_2 = tube side exit temperature = 144°F (water)

$$P = \frac{t_2 - t_1}{T_1 - t_1} = \frac{144\text{°F} - 195\text{°F}}{65\text{°F} - 195\text{°F}} = 0.39$$

$$R = \frac{T_1 - T_2}{t_2 - t_1} = \frac{65\text{°F} - 100\text{°F}}{144\text{°F} - 195\text{°F}} = 0.69$$

From the chart corresponding to $P = 0.39$ and $R = 0.69$, the MTD correction factor is $F = \boxed{0.965.}$

The answer is D.

2.9. The heat exchanger area is determined using the following equation.

$$A = \frac{q}{UF\Delta T_{\text{LMTD}}} = \frac{257{,}040 \frac{\text{Btu}}{\text{hr}}}{\left(95 \frac{\text{Btu}}{\text{hr-ft}^2\text{-°F}}\right)(0.965)(86.75\text{°F})} = \boxed{32.32 \text{ ft}^2}$$

The answer is C.

2.10. For glycol,

$$C_g = \dot{m}_g c_{pg} = \left(12{,}000 \frac{\text{lbm}}{\text{hr}}\right)\left(0.612 \frac{\text{Btu}}{\text{lbm-°F}}\right) = 7344 \text{ Btu/hr-°F}$$

For water,

$$C_w = \dot{m}_w c_{pw} = \left(5000 \frac{\text{lbm}}{\text{hr}}\right)\left(0.9994 \frac{\text{Btu}}{\text{lbm-°F}}\right) = 4997 \text{ Btu/hr-°F}$$

$C_{\min} = 4997$ Btu/hr-°F and $C_{\max} = 7344$ Btu/hr-°F

$$C = \frac{C_{\min}}{C_{\max}} = \frac{4997 \frac{\text{Btu}}{\text{hr-°F}}}{7344 \frac{\text{Btu}}{\text{hr-°F}}} = 0.6804$$

The number of transfer units is

$$\text{NTU} = N = \frac{U_i A_i}{C_{\text{min}}} = \frac{\left(188.7\ \frac{\text{Btu}}{\text{hr-ft}^2\text{-}^\circ\text{F}}\right)(15.66\ \text{ft}^2)}{4997\ \frac{\text{Btu}}{\text{hr-}^\circ\text{F}}}$$
$$= 0.5914$$

For a counterflow, double-pipe heat exchanger, the effectiveness is determined using the following equation.

$$\varepsilon = \frac{1 - \exp\big(-N(1-C)\big)}{1 - C\exp\big(-N(1-C)\big)}$$
$$= \frac{1 - \exp\big(-(0.5914)(1-0.6804)\big)}{1 - 0.6804\exp\big(-(0.5914)(1-0.6804)\big)}$$
$$= \boxed{0.3943\quad(39.43\%)}$$

The answer is B.

2.11. The heat exchanger effectiveness is

$$\varepsilon = \frac{q_{\text{actual}}}{q_{\text{max}}}$$
$$q_{\text{max}} = C_{\text{min}}\left(T_{h,\text{inlet}} - T_{c,\text{inlet}}\right)$$
$$= \left(4997\ \frac{\text{Btu}}{\text{hr-}^\circ\text{F}}\right)(195^\circ\text{F} - 50^\circ\text{F})$$
$$= 724{,}565\ \text{Btu/hr}$$
$$q_{\text{actual}} = \varepsilon q_{\text{max}} = (0.3943)\left(724{,}565\ \frac{\text{Btu}}{\text{hr}}\right)$$
$$= \boxed{285{,}696\ \text{Btu/hr}}$$

The answer is D.

2.12. $$q = \dot{m}_g c_{pg}\Delta T_g = C_g \Delta T_g$$

$$\Delta T_g = \frac{q}{C_g} = \frac{285{,}696\ \frac{\text{Btu}}{\text{hr}}}{7344\ \frac{\text{Btu}}{\text{hr-}^\circ\text{F}}} = 38.90^\circ\text{F}$$
$$T_{\text{exit},g} = T_{\text{inlet},g} + \Delta T_g = 50^\circ\text{F} + 39^\circ\text{F}$$
$$= \boxed{89^\circ\text{F}}$$

The answer is C.

2.13. $$q = \dot{m}_w c_{pw}\Delta T_w = C_w \Delta T_w$$

$$\Delta T_w = \frac{q}{C_w} = \frac{285{,}696\ \frac{\text{Btu}}{\text{hr}}}{4997\ \frac{\text{Btu}}{\text{hr-}^\circ\text{F}}} = 57.17^\circ\text{F}$$
$$T_{\text{exit},w} = T_{\text{inlet},w} - \Delta T_w = 195^\circ\text{F} - 57^\circ\text{F}$$
$$= \boxed{138^\circ\text{F}}$$

The answer is B.

SOLUTION 3

The temperature of the gas stream is T_∞. The thermocouple reading is T_t, and the surface temperature of the pipe is T_s. A is the surface area of the thermocouple. Since $T_s > T_\infty$, the heat balance at steady state results in the following.

heat gained by the thermocouple due to radiation from the pipe surface
= heat transferred to the gas by convection

$$\sigma A\varepsilon(T_s^4 - T_t^4) = hA(T_t - T_\infty)$$
$$T_t - T_\infty = \left(\frac{\sigma\varepsilon}{h}\right)(T_s^4 - T_t^4)$$
$$= \left(\frac{\left(0.1714\times 10^{-8}\ \frac{\text{Btu}}{\text{hr-ft}^2\text{-}^\circ\text{R}^4}\right)(0.5)}{40\ \frac{\text{Btu}}{\text{hr-ft}^2\text{-}^\circ\text{R}}}\right)$$
$$\times\left((1160^\circ\text{R})^4 - (760^\circ\text{R})^4\right)$$
$$= 31.64^\circ\text{R}$$

Therefore, the temperature of the gas is

$$T_\infty = T_t - 31.64^\circ\text{R} = 760^\circ\text{R} - 31.64^\circ\text{R}$$
$$= 728.4^\circ\text{R} - 460^\circ$$
$$= \boxed{268.4^\circ\text{F}}$$

The answer is A.

Mass Transfer

PROBLEM 1

A continuous dryer is designed to produce 20 000 kg of product containing 5% water in a 24 h period. The product enters the dryer with 35% water. The air used for drying has an inlet temperature of 20°C and a relative humidity of 45%, and it is preheated to 55°C before it is fed to the dryer. The air leaves the dryer fully saturated at 32°C.

1.1. What is the mass flow rate of dry (0% moisture) solid?

(A) 792 kg/h
(B) 833 kg/h
(C) 896 kg/h
(D) 948 kg/h

1.2. What is the feed rate of the wet solid (35% moisture) to the dryer?

(A) 1055 kg/h
(B) 1218 kg/h
(C) 1281 kg/h
(D) 1346 kg/h

1.3. What is the rate of removal of water from the wet solid in the dryer?

(A) 384 kg/h
(B) 427 kg/h
(C) 483 kg/h
(D) 548 kg/h

1.4. What is the moisture content of the air entering the dryer?

(A) 0.0048 kg water/kg dry air
(B) 0.0052 kg water/kg dry air
(C) 0.0058 kg water/kg dry air
(D) 0.0065 kg water/kg dry air

1.5. What is the relative humidity of the air entering the dryer?

(A) 5%
(B) 10%
(C) 15%
(D) 20%

1.6. What is the moisture content of the air leaving the dryer?

(A) 0.016 kg water/kg dry air
(B) 0.021 kg water/kg dry air
(C) 0.031 kg water/kg dry air
(D) 0.050 kg water/kg dry air

1.7. What volume flow rate of air must to be fed to the pre-heater?

(A) 1.6 m^3/s
(B) 2.4 m^3/s
(C) 3.1 m^3/s
(D) 3.7 m^3/s

1.8. What is the rate at which heat is added in the pre-heater?

(A) 146 kW
(B) 153 kW
(C) 161 kW
(D) 172 kW

PROBLEM 2

In a distillation column, it is desired to separate 200 lbmol/hr of a mixture of 40% (mol%) *n*-hexane and 60% *n*-heptane into a top product containing 95% *n*-hexane and a bottom product containing 90% *n*-heptane.

2.1. What is the mass flow rate of the bottom product?

(A) 10,600 lbm/hr
(B) 12,800 lbm/hr
(C) 14,400 lbm/hr
(D) 15,900 lbm/hr

2.2. What is the mass flow rate of the top product?

(A) 1700 lbm/hr
(B) 3500 lbm/hr
(C) 6100 lbm/hr
(D) 8400 lbm/hr

2.3. If a reflux ratio of 2:1 is used, what is the flow rate of the reflux stream?

(A) 120 lbmol/hr
(B) 140 lbmol/hr
(C) 290 lbmol/hr
(D) 330 lbmol/hr

2.4. What is the slope of the operating line for the rectifying section?

(A) 0.67
(B) 0.76
(C) 0.85
(D) 0.98

2.5. What is the y-intercept of the operating line for the rectifying section?

(A) 0.32
(B) 0.46
(C) 0.58
(D) 0.65

2.6. What is the slope of the operating line for the stripping section?

(A) 10
(B) 11
(C) 12
(D) 13

2.7. If the feed enters as a saturated liquid, the q-line is

(A) horizontal
(B) at an angle of 45° with the x-axis
(C) at an angle of 65° with the x-axis
(D) vertical

2.8. If the feed is 25% vaporized, what is the slope of the q-line?

(A) −5
(B) −4
(C) −3
(D) −2

2.9. If the feed is a subcooled liquid, what is the value of q?

(A) $q = 0$
(B) $q < 0$
(C) $q < 1$
(D) $q > 1$

PROBLEM 3

An evaporative cooling tower is used by a large power plant. The circulating water flow rate is 2000 gpm, and it enters the cooling tower at 100°F. The air inlet conditions are $T_{\text{dry bulb}} = 80°\text{F}$ and $T_{\text{wet bulb}} = 60°\text{F}$. The exit temperature of the air is 90°F, and the relative humidity is 90%.

3.1. What is the mass flow rate of the circulating water?

(A) 15,600 lbm/min
(B) 16,600 lbm/min
(C) 17,300 lbm/min
(D) 18,100 lbm/min

3.2. What is the volume flow rate of air required to cool the water to 80°F?

(A) 144,000 ft^3/min
(B) 158,000 ft^3/min
(C) 173,000 ft^3/min
(D) 183,000 ft^3/min

3.3. What is the flow rate of make-up water?

(A) 180 lbm/min
(B) 220 lbm/min
(C) 290 lbm/min
(D) 320 lbm/min

Mass Transfer Solutions

PROBLEM 1

1.1. The schematic for the process is shown as follows. ($\dot{m}_a$ is the mass flow rate of dry air.)

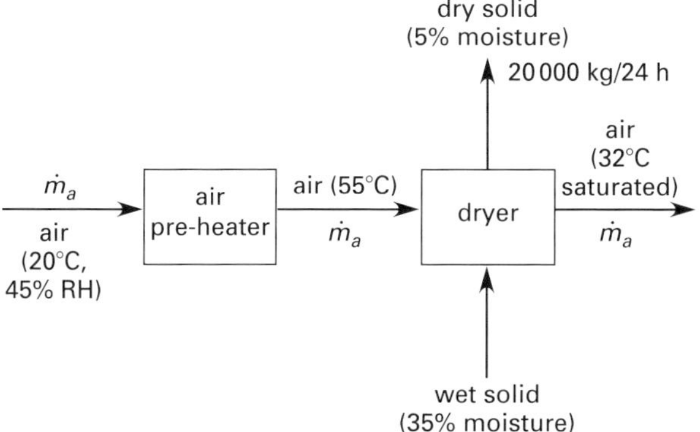

20 000 kg of product containing 5% moisture is produced in a 24 h period. This product has 95% dry solid. Therefore, the mass flow rate of dry solid is

$$\left(\frac{20\,000 \text{ kg}}{24 \text{ h}}\right)(0.95) = \boxed{792 \text{ kg/h}}$$

The answer is A.

1.2. The feed rate of wet solid to the dryer is

$$\left(792 \ \frac{\text{kg dry solid}}{\text{h}}\right)\left(\frac{1 \text{ kg wet solid}}{0.65 \text{ kg dry solid}}\right) = \boxed{1218 \text{ kg wet solid/h}}$$

The answer is B.

1.3. The rate of water removal from the wet solid is calculated as shown. The production rate of partially dry solid with 5% moisture is

$$\left(792 \ \frac{\text{kg dry solid}}{\text{h}}\right)\left(\frac{1 \text{ kg partially dry solid}}{0.95 \text{ kg dry solid}}\right) = 834 \text{ kg partially dry solid/h}$$

The rate of water removal in the dryer is

$$1218 \ \frac{\text{kg}}{\text{h}} - 834 \ \frac{\text{kg}}{\text{h}} = \boxed{384 \text{ kg/h}}$$

The answer is A.

1.4. From the psychrometric chart (SI units), at 20°C dry-bulb temperature and relative humidity of 45%, the moisture content of air entering the pre-heater is

$$\omega = 0.0065 \text{ kg water/kg dry air}$$

In the pre-heater, only the temperature of air is increased; the moisture content is unchanged. The air enters the dryer from the pre-heater. Therefore, the moisture content of air entering the dryer is

$$\omega_1 = 0.0065 \text{ kg water/kg dry air}$$

The answer is D.

1.5. Air enters the dryer at a dry-bulb temperature of 55°C and a moisture content of 0.0065 kg water/kg dry air. At this location on the psychrometric chart, the relative humidity is $\boxed{5\%.}$

The answer is A.

1.6. Air leaves the dryer fully saturated at 32°C. Under these conditions, from the psychrometric chart the moisture content is

$$\omega_2 = 0.031 \text{ kg water/kg dry air}$$

The answer is C.

1.7. The mass flow rate of dry air required is calculated as shown.

The rate of water removed from wet solid is equal to the rate of water absorbed by air.

$$384 \ \frac{\text{kg}}{\text{h}} = \dot{m}_a(\omega_2 - \omega_1)$$

$$\dot{m}_a = \frac{384 \ \frac{\text{kg water}}{\text{h}}}{\omega_2 - \omega_1} = \frac{384 \ \frac{\text{kg water}}{\text{h}}}{0.031 \ \frac{\text{kg water}}{\text{kg dry air}} - 0.0065 \ \frac{\text{kg water}}{\text{kg dry air}}} = 15\,673 \text{ kg dry air/h}$$

Note that the mass flow rate of dry air remains constant through the pre-heater and dryer.

From the psychrometric chart, at the entrance to the pre-heater (20°C dry-bulb temperature and 45% relative humidity), the specific volume is 0.84 m^3/kg dry air. Therefore, the volume flow rate of air required to be fed to the pre-heater is

$$\begin{aligned}\dot{V}_a &= \dot{m}_a v_a \\ &= \left(15\,673\ \frac{\text{kg dry air}}{\text{h}}\right)\left(0.84\ \frac{\text{m}^3}{\text{kg dry air}}\right)\left(\frac{1\ \text{h}}{3600\ \text{s}}\right) \\ &= \boxed{3.657\ \text{m}^3/\text{s}}\end{aligned}$$

The answer is D.

1.8. The rate at which heat is added in the pre-heater is

$$\dot{Q} = \dot{m}_a\left(h_2 - h_1\right)$$

h_1 = enthalpy of air entering the pre-heater (20°C, 45% relative humidity)
= 37 kJ/kg dry air

h_2 = enthalpy of air leaving the pre-heater (55°C, 5% relative humidity)
= 74 kJ/kg dry air

$$\begin{aligned}\dot{Q} &= \left(15\,673\ \frac{\text{kg dry air}}{\text{h}}\right) \\ &\quad\times\left(74\ \frac{\text{kJ}}{\text{kg dry air}} - 37\ \frac{\text{kJ}}{\text{kg dry air}}\right)\left(\frac{1\ \text{h}}{3600\ \text{s}}\right) \\ &= \boxed{161.1\ \text{kW}}\end{aligned}$$

The answer is C.

SOLUTION 2

2.1. Let F = feed, D = distillate (top product), and B = bottom product.

The overall mole balance results in

$$\begin{aligned}F &= B + D \\ 200\ \frac{\text{lbmol}}{\text{hr}} &= B + D \\ D &= 200\ \frac{\text{lbmol}}{\text{hr}} - B\end{aligned}$$

The mole balance for n-hexane results in

$$\begin{aligned}&Fx_F = Bx_B + Dx_D \\ &\left(200\ \frac{\text{lbmol}}{\text{hr}}\right)\left(0.4\ \frac{\text{lbmol hexane}}{\text{lbmol}}\right) \\ &\quad = B\left(0.1\ \frac{\text{lbmol hexane}}{\text{lbmol}}\right) \\ &\qquad + \left(200\ \frac{\text{lbmol}}{\text{hr}} - B\right)\left(0.95\ \frac{\text{lbmol hexane}}{\text{lbmol}}\right) \\ &B = 129.4\ \text{lbmol/hr}\end{aligned}$$

The molar flow rate of heptane in the bottom product is

$$\begin{aligned}&\left(0.90\ \frac{\text{lbmol heptane}}{\text{lbmol}}\right)\left(129.4\ \frac{\text{lbmol}}{\text{hr}}\right) \\ &\quad = 116.5\ \text{lbmol heptane/hr}\end{aligned}$$

The molar flow rate of hexane in the bottom product is

$$129.4\ \frac{\text{lbmol}}{\text{hr}} - 116.5\ \frac{\text{lbmol}}{\text{hr}} = 12.9\ \text{lbmol/hr}$$

The mass flow rate of heptane in the bottom product is

$$\begin{aligned}&(\text{molar flow rate})(\text{molecular weight}) \\ &\quad = \left(116.5\ \frac{\text{lbmol}}{\text{hr}}\right)\left(100\ \frac{\text{lbm}}{\text{lbmol}}\right) \\ &\quad = 11{,}650\ \text{lbm/hr}\end{aligned}$$

The mass flow rate of hexane in the bottom product is

$$\begin{aligned}&(\text{molar flow rate})(\text{molecular weight}) \\ &\quad = \left(12.9\ \frac{\text{lbmol}}{\text{hr}}\right)\left(86\ \frac{\text{lbm}}{\text{lbmol}}\right) \\ &\quad = 1109\ \text{lbm/hr}\end{aligned}$$

The total mass flow rate of the bottom product is

$$11{,}650\ \frac{\text{lbm}}{\text{hr}} + 1109\ \frac{\text{lbm}}{\text{hr}} = \boxed{12{,}759\ \text{lbm/hr}}$$

The answer is B.

2.2. $D = F - B = 200\ \frac{\text{lbmol}}{\text{hr}} - 129.4\ \frac{\text{lbmol}}{\text{hr}} = 70.6\ \text{lbmol/hr}$

The molar flow rate of hexane in the top product is

$$\begin{aligned}&\left(0.95\ \frac{\text{lbmol hexane}}{\text{lbmol}}\right)\left(70.6\ \frac{\text{lbmol}}{\text{hr}}\right) \\ &\quad = 67.1\ \text{lbmol hexane/hr}\end{aligned}$$

The molar flow rate of heptane in the top product is

$$70.6\ \frac{\text{lbmol}}{\text{hr}} - 67.1\ \frac{\text{lbmol}}{\text{hr}} = 3.5\ \text{lbmol/hr}$$

The mass flow rate of hexane in the top product is

$$\begin{aligned}&(\text{molar flow rate})(\text{molecular weight})\\&= \left(67.1\ \frac{\text{lbmol}}{\text{hr}}\right)\left(86\ \frac{\text{lbm}}{\text{lbmol}}\right)\\&= 5771\ \text{lbm/hr}\end{aligned}$$

The mass flow rate of heptane in the top product is

$$\begin{aligned}&(\text{molar flow rate})(\text{molecular weight})\\&= \left(3.5\ \frac{\text{lbmol}}{\text{hr}}\right)\left(100\ \frac{\text{lbm}}{\text{lbmol}}\right)\\&= 350\ \text{lbm/hr}\end{aligned}$$

The total mass flow rate of the top product is

$$5771\ \frac{\text{lbm}}{\text{hr}} + 350\ \frac{\text{lbm}}{\text{hr}} = \boxed{6121\ \text{lbm/hr}}$$

The answer is C.

2.3. The reflux ratio is

$$R = \frac{L}{D}$$

The molar flow rate of the reflux stream is

$$\begin{aligned}L = RD = 2D &= (2)\left(70.6\ \frac{\text{lbmol}}{\text{hr}}\right)\\&= \boxed{141.2\ \text{lbmol/hr}}\end{aligned}$$

The answer is B.

2.4. The equation for the operating line for the rectifying section is

$$y = \left(\frac{L}{L+D}\right)x + \frac{Dx_\text{D}}{L+D}$$

The slope is

$$\begin{aligned}\frac{L}{L+D} &= \frac{\frac{L}{D}}{\frac{L}{D}+1} = \frac{R}{R+1} = \frac{2}{2+1}\\&= \boxed{0.67}\end{aligned}$$

The answer is A.

2.5. The y-intercept of the operating line for the rectifying section is

$$\begin{aligned}\frac{Dx_\text{D}}{L+D} &= \frac{x_\text{D}}{\frac{L}{D}+1} = \frac{x_\text{D}}{R+1} = \frac{0.95}{2+1}\\&= \boxed{0.32}\end{aligned}$$

The answer is A.

2.6. The equation for the operating line for the stripping section is

$$y = \left(\frac{L}{L-B}\right)x - \frac{Bx_\text{B}}{L-B}$$

The slope is

$$\frac{L}{L-B} = \frac{141.2\ \frac{\text{lbmol}}{\text{hr}}}{141.2\ \frac{\text{lbmol}}{\text{hr}} - 129.4\ \frac{\text{lbmol}}{\text{hr}}} = \boxed{11.97}$$

The answer is C.

2.7. The equation for the q-line or feed line is

$$y = \left(\frac{q}{q-1}\right)x - \frac{x_F}{q-1}$$

q is the ratio of heat required to vaporize 1 mol of feed from its state to the enthalpy of vaporization, λ. Therefore,

$$q = \frac{h_G - h_F}{\lambda}$$

For saturated liquid,

$$\begin{aligned}h_G - h_F &= \lambda\\q &= \frac{\lambda}{\lambda} = 1\end{aligned}$$

For saturated liquid, the slope of the q-line is

$$\frac{q}{q-1} = \infty$$

$\boxed{\text{Therefore, the } q\text{-line is vertical.}}$

The answer is D.

2.8. If the feed is 25% vaporized,

$$q = \frac{h_\text{G} - h_\text{F}}{\lambda} = \frac{0.75\lambda}{\lambda} = 0.75$$

The slope of q-line is

$$\frac{q}{q-1} = \frac{0.75}{0.75-1} = \boxed{-3}$$

The answer is C.

2.9. If the feed is a subcooled liquid,

$$\begin{aligned}h_\text{G} - h_\text{F} &= c_p(T_b - T_\text{F}) + \lambda\\q &= \frac{h_\text{G} - h_\text{F}}{\lambda} = \frac{c_p(T_b - T_\text{F}) + \lambda}{\lambda}\\&= 1 + \frac{c_p(T_b - T_\text{F})}{\lambda}\end{aligned}$$

$$\boxed{q > 1}$$

The answer is D.

SOLUTION 3

3.1. The schematic for the cooling tower is shown.

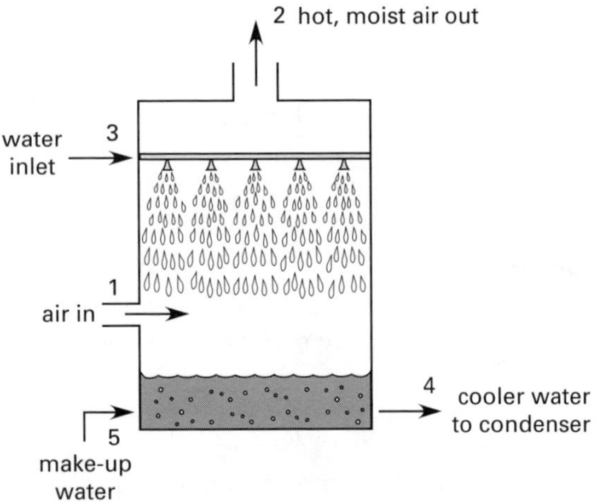

$$\dot{m}_a = \text{mass flow rate of dry air}$$
$$\dot{m}_w = \text{mass flow rate of circulating water}$$

The volume flow rate of water is

$$\dot{V}_w = \left(2000\ \frac{\text{gal}}{\text{min}}\right)\left(0.1337\ \frac{\text{ft}^3}{\text{gal}}\right) = 267.4\ \text{ft}^3/\text{min}$$

From steam tables, the specific volume of water at 100°F is

$$v_w = v_f \text{ at } 100°\text{F} = 0.0161\ \text{ft}^3/\text{lbm}$$

$$\dot{m}_w = \frac{\dot{V}_w}{v_w} = \frac{267.4\ \frac{\text{ft}^3}{\text{min}}}{0.0161\ \frac{\text{ft}^3}{\text{lbm}}}$$

$$= \boxed{16{,}609\ \text{lbm water/min}}$$

The answer is B.

3.2. $\dot{m}_m$ = mass flow rate of make-up water

The energy balance around the cooling tower is

$$\text{energy in} = \text{energy out}$$

$$\dot{m}_a h_1 + \dot{m}_w h_3 + \dot{m}_m h_5 = \dot{m}_a h_2 + \dot{m}_w h_4 \qquad \text{[I]}$$

The mass balance for water results in

$$\dot{m}_m = \dot{m}_a(\omega_2 - \omega_1)$$

Substituting into Eq. I and solving for $\dot{m}_a$,

$$\dot{m}_a = \frac{\dot{m}_w(h_3 - h_4)}{h_2 - h_1 - h_5(\omega_2 - \omega_1)} \qquad \text{[II]}$$

From steam tables,

$$h_4 = h_5 = h_f \text{ at } 80°\text{F} = 48.09\ \text{Btu/lbm water}$$
$$h_3 = h_f \text{ at } 100°\text{F} = 68.05\ \text{Btu/lbm water}$$

From the psychrometric chart the following data is obtained.

Air enters the cooling tower at a dry-bulb temperature of 80°F and a wet-bulb temperature of 60°F. Under these conditions,

$$h_1 = 27\ \text{Btu/lbm dry air}$$
$$\omega_1 = 0.0065\ \text{lbm water/lbm dry air}$$
$$v_1 = 13.75\ \text{ft}^3/\text{lbm dry air}$$

Air leaves the cooling tower at a dry-bulb temperature of 90°F and 90% relative humidity.

Under these conditions,

$$h_2 = 53\ \text{Btu/lbm dry air}$$
$$\omega_2 = 0.028\ \text{lbm water/lbm dry air}$$

Substituting the preceding data into Eq. II,

$$\dot{m}_a = \frac{\left(16{,}609\ \frac{\text{lbm water}}{\text{min}}\right) \times \left(68.05\ \frac{\text{Btu}}{\text{lbm water}} - 48.09\ \frac{\text{Btu}}{\text{lbm water}}\right)}{53\ \frac{\text{Btu}}{\text{lbm dry air}} - 27\ \frac{\text{Btu}}{\text{lbm dry air}} - \left(48.09\ \frac{\text{Btu}}{\text{lbm water}}\right) \times \left(0.028\ \frac{\text{lbm water}}{\text{lbm dry air}} - 0.0065\ \frac{\text{lbm water}}{\text{lbm dry air}}\right)}$$

$$= 13{,}279\ \text{lbm dry air/min}$$

The volume flow rate of air required is

$$\dot{V}_a = \dot{m}_a v_1$$
$$= \left(13{,}729\ \frac{\text{lbm dry air}}{\text{min}}\right)\left(13.75\ \frac{\text{ft}^3}{\text{lbm dry air}}\right)$$
$$= \boxed{182{,}586\ \text{ft}^3/\text{min}}$$

The answer is D.

3.3. The make-up water flow rate required is

$$\dot{m}_m = \dot{m}_a(\omega_2 - \omega_1)$$
$$= \left(13{,}279\ \frac{\text{lbm dry air}}{\text{min}}\right) \times \left(0.028\ \frac{\text{lbm water}}{\text{lbm dry air}} - 0.0065\ \frac{\text{lbm water}}{\text{lbm dry air}}\right)$$
$$= \boxed{286\ \text{lbm water/min}}$$

The answer is C.

Kinetics

PROBLEM 1

The decomposition of a certain gas proceeds according to a second-order reaction as follows.

$$2\,\mathrm{A}(g) \longrightarrow 2\,\mathrm{R}(g) + \mathrm{S}(g)$$

The reverse reaction is negligible. At 950°C, the rate constant, k, is 1200 $\mathrm{cm^3/mol{\cdot}s}$. The initial reaction mixture consists of pure A. The temperature is maintained at 950°C in a batch reactor, and the pressure is constant at 1 atm.

1.1. What is the design equation for this reactor?

(A) $t = \dfrac{N_{\mathrm{A0}}}{V}\displaystyle\int_0^{X_\mathrm{A}} \frac{dX_\mathrm{A}}{-r_\mathrm{A}}$

(B) $t = C_{\mathrm{A0}}\displaystyle\int_0^{X_\mathrm{A}} \frac{dX_\mathrm{A}}{(-r_\mathrm{A})^2}$

(C) $t = \displaystyle\int_0^{C_\mathrm{A}} \frac{dC_\mathrm{A}}{r_\mathrm{A}}$

(D) $t = N_{\mathrm{A0}}\displaystyle\int_0^{X_\mathrm{A}} \frac{dX_\mathrm{A}}{(-r_\mathrm{A})V}$

1.2. What is the rate expression for the reaction?

(A) $-r_\mathrm{A} = kC_{\mathrm{A0}}^2(1-X_\mathrm{A})^2$

(B) $-r_\mathrm{A} = \dfrac{kC_{\mathrm{A0}}^2(1-X_\mathrm{A})^2}{(1+\varepsilon_\mathrm{A}X_\mathrm{A})^2}$

(C) $-r_\mathrm{A} = \dfrac{kC_{\mathrm{A0}}(1-X_\mathrm{A})}{1+\varepsilon_\mathrm{A}X_\mathrm{A}}$

(D) $-r_\mathrm{A} = \dfrac{kC_{\mathrm{A0}}^2(1-X_\mathrm{A}^2)}{1+\varepsilon_\mathrm{A}^2X_\mathrm{A}^2}$

1.3. Which equation can be used to calculate the time required for the proposed conversion?

(A) $t = \dfrac{1}{kC_{\mathrm{A0}}^2}\displaystyle\int_0^{X_\mathrm{A}} \frac{(1+\varepsilon_\mathrm{A}X_\mathrm{A})\,dX_\mathrm{A}}{(1-X_\mathrm{A})^2}$

(B) $t = \dfrac{1}{kC_{\mathrm{A0}}^2}\displaystyle\int_0^{X_\mathrm{A}} \frac{dX_\mathrm{A}}{(1-X_\mathrm{A})^2}$

(C) $t = \dfrac{1}{kC_{\mathrm{A0}}}\displaystyle\int_0^{X_\mathrm{A}} \frac{(1+\varepsilon_\mathrm{A}X_\mathrm{A})\,dX_\mathrm{A}}{(1-X_\mathrm{A})^2}$

(D) $t = \dfrac{1}{kC_{\mathrm{A0}}}\displaystyle\int_0^{X_\mathrm{A}} \frac{(1+\varepsilon_\mathrm{A}X_\mathrm{A})^2\,dX_\mathrm{A}}{(1-X_\mathrm{A})^2}$

1.4. What is the initial concentration of the reaction mixture?

(A) 0.01 mol/L
(B) 0.02 mol/L
(C) 0.03 mol/L
(D) 0.04 mol/L

1.5. What is the fractional change in volume, ε_A, for complete conversion of A?

(A) 0.30
(B) 0.40
(C) 0.50
(D) 0.60

1.6. What is the time required for 90% conversion of A?

(A) 17 min
(B) 23 min
(C) 32 min
(D) 41 min

1.7. What is the final concentration of A after 90% conversion?

(A) 0.0003 mol/L
(B) 0.0005 mol/L
(C) 0.0007 mol/L
(D) 0.0009 mol/L

1.8. What is the final concentration of S?

(A) 0.003 mol/L
(B) 0.006 mol/L
(C) 0.008 mol/L
(D) 0.010 mol/L

PROBLEM 2

In a liquid-phase reaction, reactant A is converted to product B in the presence of catalyst C.

$$A + C \longrightarrow B + C$$

The rate expression for this reaction is

$$-r_A = kC_A C_C$$

The rate constant is $k = 1.20 \times 10^{-5} \text{m}^3/\text{kmol·s}$, and the concentration of the catalyst can be assumed to be constant throughout the reactor. The liquid feed rate is 0.001 m^3/s, and the feed consists of 90 $kmol/m^3$ of A and 10 $kmol/m^3$ of C.

2.1. What is the molar feed rate of A?

(A) 0.03 kmol/s
(B) 0.06 kmol/s
(C) 0.07 kmol/s
(D) 0.09 kmol/s

2.2. If the molecular weight of B is 40 kg/kmol, what is the production rate of B in kg/s for 60% conversion of A?

(A) 1.1 kg/s
(B) 1.4 kg/s
(C) 1.8 kg/s
(D) 2.2 kg/s

2.3. What is the reaction rate when the conversion is 50%?

(A) 0.001 $kmol/m^3·s$
(B) 0.003 $kmol/m^3·s$
(C) 0.005 $kmol/m^3·s$
(D) 0.008 $kmol/m^3·s$

2.4. For 50% conversion, what volume of a plug flow reactor would be required?

(A) 4.9 m^3
(B) 5.8 m^3
(C) 6.5 m^3
(D) 7.1 m^3

2.5. What is the time required for 50% conversion in a plug flow reactor?

(A) 1.6 h
(B) 1.9 h
(C) 2.2 h
(D) 2.7 h

2.6. What is the volume of a mixed-flow reactor required for 50% conversion?

(A) 7.1 m^3
(B) 8.3 m^3
(C) 9.1 m^3
(D) 9.9 m^3

2.7. What is the time required for 50% conversion in a mixed-flow reactor?

(A) 1.5 h
(B) 1.9 h
(C) 2.3 h
(D) 2.7 h

PROBLEM 3

The following information on the rate constant is available for a first-order reaction.

temperature (K)	rate constant, k (s^{-1})
310	0.0004
340	0.0072

3.1. What is the activation energy for this reaction?

(A) 63 400 J/mol
(B) 72 200 J/mol
(C) 84 400 J/mol
(D) 97 300 J/mol

3.2. What is the frequency factor, A?

(A) $3.2 \times 10^{10}\ s^{-1}$
(B) $4.8 \times 10^{10}\ s^{-1}$
(C) $6.7 \times 10^{10}\ s^{-1}$
(D) $8.2 \times 10^{10}\ s^{-1}$

3.3. What temperature increase is necessary for the reaction rate to double from its value at 200°C?

(A) 11°C
(B) 16°C
(C) 24°C
(D) 29°C

Kinetics Solutions

SOLUTION 1

1.1. The general design equation for a batch reactor is

$$\boxed{t = N_{A0}\int_0^{X_A}\frac{dX_A}{(-r_A)V}}$$

This is a gas phase reaction where the mols of the reactants and products are not equal. The reaction volume changes with conversion. Therefore, the volume term cannot be removed from the integral sign.

The answer is D.

1.2. For a variable-volume system,

$$C_A = \frac{N_A}{V} = \frac{N_{A0}(1-X_A)}{V_0(1+\varepsilon_A X_A)} = \frac{C_{A0}(1-X_A)}{1+\varepsilon_A X_A}$$

For a second-order reaction, the rate expression is

$$\boxed{-r_A = kC_A^2 = \frac{kC_{A0}^2(1-X_A)^2}{(1+\varepsilon_A X_A)^2}}$$

The answer is B.

1.3. Substituting $V = V_0(1+\varepsilon_A X_A)$ and the expression for $-r_A$ in the design equation in Prob. 1.1,

$$\begin{aligned} t &= N_{A0}\int_0^{X_A}\frac{dX_A(1+\varepsilon_A X_A)^2}{kC_{A0}^2(1-X_A)^2 V_0(1+\varepsilon_A X_A)} \\ &= \frac{N_{A0}}{V_0}\int_0^{X_A}\frac{(1+\varepsilon_A X_A)\,dX_A}{kC_{A0}^2(1-X_A)^2} \\ &= \frac{C_{A0}}{kC_{A0}^2}\int_0^{X_A}\frac{(1+\varepsilon_A X_A)\,dX_A}{(1-X_A)^2} \\ &= \boxed{\frac{1}{kC_{A0}}\int_0^{X_A}\frac{(1+\varepsilon_A X_A)\,dX_A}{(1-X_A)^2}} \end{aligned}$$

The answer is C.

1.4.

$$\begin{aligned} C_{A0} &= \frac{p}{RT} = \frac{1\text{ atm}}{\left(0.0821\ \frac{\text{L·atm}}{\text{mol·K}}\right)(1228\text{K})} \\ &= \boxed{0.0100\text{ mol/L}} \end{aligned}$$

The answer is A.

1.5.

$$\varepsilon_A = \frac{V_{X_A=1} - V_{X_A=0}}{V_{X_A=0}} = \frac{3-2}{2} = \boxed{0.5}$$

The answer is C.

1.6. From any table of integrals,

$$\int_0^{X_A}\frac{(1+\varepsilon_A X_A)\,dX_A}{(1-X_A)^2} = \frac{(1+\varepsilon_A)X_A}{1-X_A} + \varepsilon_A\ln(1-X_A)$$

Therefore, using the results from Prob. 1.3,

$$t = \left(\frac{1}{kC_{A0}}\right)\left(\frac{(1+\varepsilon_A)X_A}{1-X_A} + \varepsilon_A\ln(1-X_A)\right)$$

$$k = \left(1200\ \frac{\text{cm}^3}{\text{mol·s}}\right)\left(\frac{1\text{ L}}{1000\text{ cm}^3}\right) = 1.2\text{ L/mol·s}$$

$$X_A = 0.9$$

Substituting for k, X_A, C_{A0}, and ε_A into the expression for t,

$$\begin{aligned} t &= \left(\frac{1}{\left(1.2\ \frac{\text{L}}{\text{mol·s}}\right)\left(0.01\ \frac{\text{mol}}{\text{L}}\right)}\right) \\ &\quad\times\left(\frac{(1+0.5)(0.9)}{1-0.9} + 0.5\ln(1-0.9)\right)\left(\frac{1\text{ min}}{60\text{ s}}\right) \\ &= \boxed{17.15\text{ min}} \end{aligned}$$

The answer is A.

1.7. The final concentration of A is given by

$$\begin{aligned} C_A &= \frac{C_{A0}(1-X_A)}{1+\varepsilon_A X_A} = \frac{\left(0.0100\ \frac{\text{mol}}{\text{L}}\right)(1-0.9)}{1+(0.5)(0.9)} \\ &= \boxed{0.00069\text{ mol/L}} \end{aligned}$$

The answer is C.

1.8. The final concentration of S is determined as shown.

$$\text{mol of S formed} = \frac{\text{mol of A reacted}}{2}$$

$$N_S = \frac{N_{A0}X_A}{2}$$

$$C_S = \frac{N_S}{V} = \frac{N_{A0}X_A}{2V}$$

Since $V = V_0(1+\varepsilon_A X_A)$,

$$C_S = \frac{N_{A0}X_A}{2V} = \frac{N_{A0}X_A}{2V_0(1+\varepsilon_A X_A)} = \frac{C_{A0}X_A}{(2)(1+\varepsilon_A X_A)}$$

Note that $N_{A0}/V_0 = C_{A0}$.

$$C_S = \frac{C_{A0}X_A}{(2)(1+\varepsilon_A X_A)} = \frac{\left(0.0100\ \frac{\text{mol}}{\text{L}}\right)(0.9)}{(2)(1+(0.5)(0.9))}$$

$$= \boxed{0.0031\ \text{mol/L}}$$

The answer is A.

SOLUTION 2

2.1. The molar feed rate of A is given by

$$F_{A0} = C_{A0}\text{v}_0 = \left(90\ \frac{\text{kmol}}{\text{m}^3}\right)\left(0.001\ \frac{\text{m}^3}{\text{s}}\right)$$

$$= \boxed{0.09\ \text{kmol/s}}$$

The answer is D.

2.2. From the stoichiometry of the reaction,

$$\text{mol of B formed} = \text{mol of A reacted}$$

$$F_B = F_{A0}X_A = \left(0.09\ \frac{\text{kmol}}{\text{s}}\right)(0.60) = 0.054\ \text{kmol/s}$$

The production rate of B is

$$F_B M_B = \left(0.054\ \frac{\text{kmol}}{\text{s}}\right)\left(40\ \frac{\text{kg}}{\text{kmol}}\right) = \boxed{2.16\ \text{kg/s}}$$

The answer is D.

2.3. The reaction rate is given by

$$-r_A = kC_A C_C = kC_{A0}(1-X_A)C_C$$

$$= \left(1.20\times10^{-5}\ \frac{\text{m}^3}{\text{kmol·s}}\right)\left(90\ \frac{\text{kmol}}{\text{m}^3}\right) \times (1-0.5)\left(10\ \frac{\text{kmol}}{\text{m}^3}\right)$$

$$= \boxed{0.0054\ \text{kmol/m}^3\text{·s}}$$

The answer is C.

2.4. The design equation for a plug flow reactor is

$$V = F_{A0}\int_0^{X_A} \frac{dX_A}{-r_A}$$

$$-r_A = kC_A C_C = kC_{A0}(1-X_A)C_C$$

$$V = F_{A0}\int_0^{X_A} \frac{dX_A}{kC_{A0}(1-X_A)C_C}$$

The concentration of the catalyst, C_C, is constant throughout the reactor. k and C_{A0} are also constant. Therefore,

$$V = \frac{F_{A0}}{kC_{A0}C_C}\int_0^{X_A}\frac{dX_A}{1-X_A} = \left(\frac{F_{A0}}{kC_{A0}C_C}\right)\left(-\ln(1-X_A)\right)_0^{0.5} = \frac{0.6931F_{A0}}{kC_{A0}C_C}$$

$$\frac{F_{A0}}{C_{A0}} = \text{v}_0 = 0.001\ \text{m}^3/\text{s}$$

Substituting into the equation for V,

$$V = \frac{(0.6931)\left(0.001\ \frac{\text{m}^3}{\text{s}}\right)}{\left(1.20\times10^{-5}\ \frac{\text{m}^3}{\text{kmol·s}}\right)\left(10\ \frac{\text{kmol}}{\text{m}^3}\right)} = \boxed{5.776\ \text{m}^3}$$

The answer is B.

2.5. The space time, τ, for a plug flow reactor is given by

$$\tau = \frac{V}{\text{v}_0} = \left(\frac{5.776\ \text{m}^3}{0.001\ \frac{\text{m}^3}{\text{s}}}\right)\left(\frac{1\ \text{h}}{3600\ \text{s}}\right) = \boxed{1.60\ \text{h}}$$

The answer is A.

2.6. The volume of a mixed-flow reactor can be calculated using the design equation.

$$V = \frac{F_{A0}X_A}{-r_A}$$

From the solution to Prob. 2.3, when the conversion is 50%,

$$-r_A = 0.0054\ \text{kmol/m}^3\text{·s}$$

$$V = \frac{\left(0.09\ \frac{\text{kmol}}{\text{s}}\right)(0.5)}{0.0054\ \frac{\text{kmol}}{\text{m}^3\text{·s}}} = \boxed{8.333\ \text{m}^3}$$

The answer is B.

2.7. The space time, τ, for a mixed-flow reactor is given by

$$\tau = \frac{V}{v_0} = \left(\frac{8.333\ \text{m}^3}{0.001\ \frac{\text{m}^3}{\text{s}}}\right)\left(\frac{1\ \text{h}}{3600\ \text{s}}\right) = \boxed{2.31\ \text{h}}$$

The answer is C.

SOLUTION 3

3.1. The Arrhenius equation for the rate constant, k, is

$$k = Ae^{\frac{-E}{\overline{R}T}}$$

A is the frequency factor and has the same units as k; E is the activation energy in J/mol; $\overline{R}$ is the universal gas constant, 8.314 J/mol·K; and T is the absolute temperature in K.

Taking the natural log of the Arrhenius equation,

$$\ln k = \ln A - \left(\frac{E}{\overline{R}}\right)\left(\frac{1}{T}\right)$$

$$\ln k_1 = \ln A - \left(\frac{E}{\overline{R}}\right)\left(\frac{1}{T_1}\right)$$

$$\ln k_2 = \ln A - \left(\frac{E}{\overline{R}}\right)\left(\frac{1}{T_2}\right)$$

Subtracting,

$$\ln\left(\frac{k_2}{k_1}\right) = -\left(\frac{E}{\overline{R}}\right)\left(\frac{1}{T_2} - \frac{1}{T_1}\right)$$

Substituting the given data,

$$\ln\left(\frac{0.0072\ \text{s}^{-1}}{0.0004\ \text{s}^{-1}}\right) = -\left(\frac{E}{\overline{R}}\right)\left(\frac{1}{340\text{K}} - \frac{1}{310\text{K}}\right)$$

$$\frac{-E}{\overline{R}} = -10\,155\text{K}$$

$$E = (10\,155\text{K})(\overline{R})$$

$$= (10\,155\text{K})\left(8.314\ \frac{\text{J}}{\text{mol·K}}\right)$$

$$= \boxed{84\,429\ \text{J/mol}}$$

The answer is C.

3.2.

$$A = \frac{k_1}{e^{\frac{-E}{\overline{R}T_1}}} = k_1 e^{\frac{E}{\overline{R}T_1}}$$

$$= (0.0004\ \text{s}^{-1})\left(e^{\frac{84\,429\ \frac{\text{J}}{\text{mol}}}{\left(8.314\ \frac{\text{J}}{\text{mol·K}}\right)(310\text{K})}}\right)$$

$$= \boxed{6.742 \times 10^{10}\ \text{s}^{-1}}$$

The answer is C.

3.3.

$$T_1 = 200°\text{C} + 273° = 473\text{K}$$

$$\frac{k_2}{k_1} = 2$$

$$\ln\left(\frac{k_2}{k_1}\right) = -\left(\frac{E}{\overline{R}}\right)\left(\frac{1}{T_2} - \frac{1}{T_1}\right)$$

$$\ln(2) = -(10\,155\text{K})\left(\frac{1}{T_2} - \frac{1}{473\text{K}}\right)$$

Solving for T_2,

$$T_2 = 489\text{K} - 273° = 216°\text{C}$$

Therefore, the temperature increase required is

$$216°\text{C} - 200°\text{C} = \boxed{16°\text{C}}$$

The answer is B.

Plant Design

PROBLEM 1

A 3 in nominal pipe carrying steam at 200°C is to be insulated. The pipe is exposed to ambient air at 5°C. The heat transfer coefficient for steam is 2000 $W/m^2\cdot°C$, and that for ambient air is 55 $W/m^2\cdot°C$. The thermal conductivity of the insulating material is 0.05 W/m·°C. It has been estimated that each kilowatt-hour of energy loss from the pipe costs $0.001. The installed cost of insulation is $860/m^3$. The steam line is in operation 5000 h during a year and is 200 m long. The inside diameter of the pipe is 78 mm. Neglect the thickness of the pipe wall and the resistance due to the pipe wall.

1.1. What is the heat loss from the bare pipe?

(A) 260 kW
(B) 370 kW
(C) 430 kW
(D) 510 kW

1.2. If the heat loss is reduced by 95% by using insulation, what is the heat loss from the insulated pipe?

(A) 26 kW
(B) 33 kW
(C) 73 kW
(D) 82 kW

1.3. What is the insulation thickness required to reduce the heat loss by 95%?

(A) 12 mm
(B) 23 mm
(C) 36 mm
(D) 58 mm

1.4. What is the surface area of the pipe with insulation?

(A) 56 m^2
(B) 63 m^2
(C) 69 m^2
(D) 77 m^2

1.5. What is the resistance due to insulation?

(A) 0.0021°C/W
(B) 0.0036°C/W
(C) 0.0072°C/W
(D) 0.0091°C/W

1.6. If the interest rate is 8% and the useful life of the insulation is 5 yr, what is the annual cost of insulation?

(A) $380/yr
(B) $570/yr
(C) $710/yr
(D) $920/yr

1.7. What is the net annual savings due to the installation of insulation?

(A) $1400/yr
(B) $1800/yr
(C) $2100/yr
(D) $2500/yr

PROBLEM 2

15 kg/h of B (MW = 45 g/mol) are to be produced from a feed consisting of a saturated solution of A in a mixed-flow reactor. The reaction is as follows.

$$\mathrm{A} \longrightarrow \mathrm{B}$$

$$-r_\mathrm{A} = kC_\mathrm{A} \text{ and } k = 0.005\ \mathrm{s}^{-1}$$

The initial concentration of A is 0.2 mol/L, and the cost of A is $0.85/mol. The annual equivalent cost of the reactor is 120V$ where V is the reactor volume in liters. This cost includes installation, instrumentation, and other ancillary equipment. The operating cost is $25/h, and the reactor will operate 6000 h/yr. X_A is the fractional conversion of A.

2.1. What is the equation for reactor volume in terms of the molar production rate of B, F_B?

(A) $V = \dfrac{F_B(1 - X_A)}{kC_{A0}}$

(B) $V = \dfrac{F_B C_{A0}}{k(1 - X_A)}$

(C) $V = \dfrac{kF_B}{C_{A0}(1 - X_A)}$

(D) $V = \dfrac{F_B}{kC_{A0}(1 - X_A)}$

2.2. In terms of conversion of A, X_A, what is the annual equivalent cost of the reactor?

(A) $\$\left(\dfrac{10\,223}{1 - X_A}\right)/\text{yr}$

(B) $\$\left(\dfrac{11\,111}{1 - X_A}\right)/\text{yr}$

(C) $\$\left(\dfrac{14\,323}{1 - X_A}\right)/\text{yr}$

(D) $\$\left(\dfrac{16\,568}{1 - X_A}\right)/\text{yr}$

2.3. What is the annual cost of the reactant A?

(A) $\$\left(\dfrac{1.7 \times 10^6}{X_A}\right)/\text{yr}$

(B) $\$\left(\dfrac{2.3 \times 10^6}{X_A}\right)/\text{yr}$

(C) $\$\left(\dfrac{3.4 \times 10^6}{X_A}\right)/\text{yr}$

(D) $\$\left(\dfrac{4.6 \times 10^6}{X_A}\right)/\text{yr}$

2.4. What is the fractional conversion of A that minimizes the total costs?

(A) 0.66
(B) 0.78
(C) 0.86
(D) 0.93

2.5. What is the optimum reactor size?

(A) 1050 L
(B) 1190 L
(C) 1240 L
(D) 1390 L

2.6. What molar flow rate of A is required?

(A) 280 mol/h
(B) 360 mol/h
(C) 410 mol/h
(D) 490 mol/h

2.7. What is the cost of production of B?

(A) \$19/kg
(B) \$21/kg
(C) \$24/kg
(D) \$26/kg

PROBLEM 3

A process requires an 8% potassium hydroxide (KOH) solution at the rate of 1200 gal/hr. The safety of the process dictates a tolerance of ±0.2% in KOH concentration. A surge tank is used to smooth out variations in flow rate and concentrations of KOH solution before the solution is fed to the process. The inlet flow to the surge tank is 900 gal/hr of 8.4% KOH solution, and the tank contains 6000 gal of KOH solution. A level alarm will sound if the liquid level in the tank drops below 4 ft. The tank has a diameter of 10 ft. The density of the solution is 8.6 lbm/gal.

3.1. At its full capacity of 6000 gal, what is the height of the liquid in the tank?

(A) 3 ft
(B) 6 ft
(C) 8 ft
(D) 10 ft

3.2. If the level alarm did not sound, how much time would be required to empty the tank?

(A) 13 hr
(B) 15 hr
(C) 18 hr
(D) 20 hr

3.3. Under the stated conditions, how much time elapses before the level alarm sounds?

(A) 12 hr
(B) 15 hr
(C) 18 hr
(D) 21 hr

Consider the following situation for Probs. 3.4–3.6. Due to the malfunction of a valve, the concentration of potassium hydroxide in the inlet stream to the surge tank drops instantaneously to 6%, but the flow rate remains the same. The contents of the tank are well mixed. The concentrations are on a mass basis.

3.4. A process monitoring alarm will sound if the KOH concentration in the outlet stream drops below 7.8%. What is the time required for this alarm to sound?

(A) 20 min
(B) 25 min
(C) 32 min
(D) 42 min

3.5. How much time is required for the KOH concentration in the outlet stream to reach 7% (assuming the alarm does not sound)?

(A) 4.1 hr
(B) 5.2 hr
(C) 6.1 hr
(D) 7.3 hr

3.6. What is the concentration of KOH in the tank after 6 hr?

(A) 4.1%
(B) 5.1%
(C) 5.9%
(D) 6.7%

PROBLEM 4

A plant effluent stream contains component W that is harmful to the environment. The effluent stream is oxidized to produce less harmful products according to the following reaction.

$$W + O_2 \longrightarrow \text{products}$$

The rate equation is of the form

$$-r_W = kC_WC_{O_2}$$

The reaction takes place in a large tank that is well aerated and mixed. The effluent stream flow rate is 1500 gal/hr, and the concentration of W in the stream is 1.5×10^{-4} lbmol/gal. The oxygen concentration is maintained constant at 2.5×10^{-6} lbmol/gal. The reaction rate constant, k, is 43,500 gal/lbmol-hr.

4.1. What is the design equation for the reactor?

(A) $V = \dfrac{F_{W0}(1 - X_W)}{-r_W}$

(B) $V = \dfrac{F_{W0}X_W}{-r_W}$

(C) $V = \dfrac{F_WX_W}{r_W}$

(D) $V = \dfrac{F_{W0}(-r_W)}{X_W}$

4.2. What is the molar flow rate of W entering the reactor?

(A) 0.23 lbmol/hr
(B) 0.31 lbmol/hr
(C) 0.42 lbmol/hr
(D) 0.56 lbmol/hr

4.3. For 97% conversion, what is the concentration of W leaving the reactor?

(A) 1.5×10^{-6} lbmol/gal
(B) 3.1×10^{-6} lbmol/gal
(C) 4.5×10^{-6} lbmol/gal
(D) 6.3×10^{-6} lbmol/gal

4.4. What is the volume of the reactor required for 97% conversion of W?

(A) 446,000 gal
(B) 468,000 gal
(C) 481,000 gal
(D) 492,000 gal

4.5. If the effluent stream flow rate increases to 3500 gal/hr, what conversion will be achieved in the reactor?

(A) 81%
(B) 85%
(C) 89%
(D) 93%

For Probs. 4.6 through 4.9, assume a second tank is added in series to the present tank to handle the increased flow rate as well to achieve an overall conversion of 99% to meet new regulations. The new tank will operate at the same temperature and oxygen concentration as the present tank.

4.6. What is the percentage of the conversion required in the second tank?

(A) 85%
(B) 89%
(C) 93%
(D) 97%

4.7. What is the volume of the second tank?

(A) 165,000 gal
(B) 185,000 gal
(C) 205,000 gal
(D) 225,000 gal

4.8. What is the molar flow rate of W leaving the second tank?

(A) 2.3×10^{-3} lbmol/hr
(B) 3.8×10^{-3} lbmol/hr
(C) 4.3×10^{-3} lbmol/hr
(D) 5.3×10^{-3} lbmol/hr

4.9. Under the stated conditions, the reaction is

(A) second order
(B) first order
(C) zero order
(D) pseudo-first order

Plant Design Solutions

SOLUTION 1

1.1. The heat loss from bare pipe is calculated using the following equation.

$$q_1 = \frac{\Delta T_{\text{overall}}}{\frac{1}{h_i A} + \frac{1}{h_o A}}$$

$$\Delta T_{\text{overall}} = 200^\circ\text{C} - 5^\circ\text{C} = 195^\circ\text{C}$$

The inside heat transfer coefficient is $h_i = 2000\ \text{W/m}^2{\cdot}^\circ\text{C}$. The outside heat transfer coefficient is $h_o = 55\ \text{W/m}^2{\cdot}^\circ\text{C}$.

The surface area of the pipe is

$$A = \pi DL = \pi(0.078\ \text{m})(200\ \text{m}) = 49.01\ \text{m}^2$$

Therefore,

$$\begin{aligned} q_1 &= \frac{A\Delta T_{\text{overall}}}{\frac{1}{h_i} + \frac{1}{h_o}} \\ &= \left(\frac{(49.01\ \text{m}^2)(195^\circ\text{C})}{\frac{1}{2000\ \frac{\text{W}}{\text{m}^2{\cdot}^\circ\text{C}}} + \frac{1}{55\ \frac{\text{W}}{\text{m}^2{\cdot}^\circ\text{C}}}}\right)\left(\frac{1\ \text{kW}}{1000\ \text{W}}\right) \\ &= \boxed{511.6\ \text{kW}} \end{aligned}$$

The answer is D.

1.2. The heat loss from the insulated pipe is 5% of the heat loss from the base pipe.

$$q_2 = 0.05q_1 = (0.05)(511.6\ \text{kW}) = \boxed{25.6\ \text{kW}}$$

The answer is A.

1.3. The heat transfer with insulation is

$$q_2 = \frac{\Delta T_{\text{overall}}}{\frac{1}{h_i A_i} + \frac{\ln\left(\frac{r_o}{r_i}\right)}{2\pi k L} + \frac{1}{h_o A_o}} \qquad \text{[I]}$$

$$A_i = 49.01\ \text{m}^2$$

If t is the insulation thickness,

$$r_o = r_i + t = \frac{0.078\ \text{m}}{2} + t = 0.039\ \text{m} + t$$

$$\begin{aligned} A_o &= \pi D_o L = \pi(0.078\ \text{m} + 2t)(200\ \text{m}) \\ &= 49.01\ \text{m}^2 + 1257t \end{aligned} \qquad \text{[II]}$$

As an approximation,

$$A_o = A_i = 49.01\ \text{m}^2$$

Substituting into Eq. I,

$$\begin{aligned} 25\,600\ \text{W} &= \frac{195^\circ\text{C}}{\begin{array}{l}\frac{1}{\left(2000\ \frac{\text{W}}{\text{m}^2{\cdot}^\circ\text{C}}\right)(49.01\ \text{m}^2)} \\ \quad + \frac{\ln\left(\frac{0.039+t}{0.039}\right)}{2\pi\left(0.05\ \frac{\text{W}}{\text{m}{\cdot}^\circ\text{C}}\right)(200\ \text{m})} \\ \quad + \frac{1}{\left(55\ \frac{\text{W}}{\text{m}^2{\cdot}^\circ\text{C}}\right)(49.01\ \text{m}^2)}\end{array}} \qquad \text{[III]} \\ &= \frac{195^\circ\text{C}}{\begin{array}{l}0.00001\ \frac{^\circ\text{C}}{\text{W}} \\ \quad + \left(0.0159\ \frac{^\circ\text{C}}{\text{W}}\right)\ln\left(\frac{0.039+t}{0.039}\right) \\ \quad + 0.00037\ \frac{^\circ\text{C}}{\text{W}}\end{array}} \end{aligned}$$

Simplifying,

$$\ln\left(\frac{0.039+t}{0.039}\right) = 0.4552$$

$$\frac{0.039+t}{0.039} = e^{0.4552} = 1.576$$

$$t = \boxed{0.0225\ \text{m} \quad (22.5\ \text{mm})}$$

The answer is B.

1.4. From Eq. II of Prob. 1.3,

$$\begin{aligned} A_o &= 49.01\ \text{m}^2 + (1257\ \text{m})t \\ &= 49.01\ \text{m}^2 + (1257\ \text{m})(0.0225\ \text{m}) \\ &= \boxed{77.29\ \text{m}^2} \end{aligned}$$

The answer is D.

1.5. From Eq. III of Prob. 1.3, the resistance due to insulation is given by

$$R_{\text{ins}} = \left(0.0159\ \frac{^\circ\text{C}}{\text{W}}\right)\ln\left(\frac{0.039+t}{0.039}\right)$$

Since $t = 0.0225$ m,

$$\begin{aligned} R_{\text{ins}} &= \left(0.0159\ \frac{^\circ\text{C}}{\text{W}}\right)\ln\left(\frac{0.039+0.0225}{0.039}\right) \\ &= 0.0072\ ^\circ\text{C/W} \end{aligned}$$

Note that although $A_o = 77.29\ \text{m}^2$ and $A_i = 49.01\ \text{m}^2$, the use of A_i and A_o interchangeably in Eq. I of Prob. 1.3 will not affect the accuracy of the final result significantly because the resistance terms involving A_o and A_i are appreciably smaller than the resistance term involving the insulation.

The answer is C.

1.6. The volume of insulation required is

$$(77.29\ \text{m}^2)(0.0225\ \text{m}) = 1.739\ \text{m}^3$$

The installed cost of the insulation is

$$\left(\frac{\$860}{1\ \text{m}^3}\right)(1.739\ \text{m}^3) = \$1496$$

At 8% interest and with a 5 yr useful life, the annual cost of insulation is

$$(\$1496)(A/P,\ 8\%,\ 5) = (\$1496)(0.2505) = \boxed{\$375/\text{yr}}$$

The answer is A.

1.7. The reduction in heat loss is

$$0.95q = (0.95)(511.6\ \text{kW}) = 486.0\ \text{kW}$$

The annual savings due to the reduction in heat loss is

$$(486.0\ \text{kW})\left(5000\ \frac{\text{h}}{\text{yr}}\right)\left(\frac{\$0.001}{1\ \text{kW·h}}\right) = \$2430/\text{yr}$$

net annual savings = annual savings due to reduction in heat loss − annual cost of insulation

$$\$2430 - \$375 = \boxed{\$2055/\text{yr}}$$

The answer is C.

SOLUTION 2

2.1. For a first-order reaction, the volume of a mixed-flow reactor is

$$V = \frac{F_{\text{A0}}X_{\text{A}}}{-r_{\text{A}}} = \frac{F_{\text{A0}}X_{\text{A}}}{kC_{\text{A}}} = \frac{F_{\text{A0}}X_{\text{A}}}{kC_{\text{A0}}(1-X_{\text{A}})}$$

Since $F_{\text{A0}}X_{\text{A}} = F_{\text{B}}$ = rate of production of B,

$$\boxed{V = \frac{F_{\text{B}}}{kC_{\text{A0}}(1-X_{\text{A}})}}$$

The answer is D.

2.2.

$$\begin{aligned} F_{\text{B}} &= \left(15\ \frac{\text{kg}}{\text{h}}\right)\left(1000\ \frac{\text{g}}{\text{kg}}\right)\left(\frac{1\ \text{mol B}}{45\ \text{g B}}\right) \\ &= 333.33\ \text{mol B/h} \\ k &= \left(\frac{0.005}{1\ \text{s}}\right)\left(3600\ \frac{\text{s}}{\text{h}}\right) = 18/\text{h} \\ C_{\text{A0}} &= 0.2\ \text{mol/L} \end{aligned}$$

The volume of the reactor is obtained by substituting into the equation derived in the solution to Prob. 2.1.

$$\begin{aligned} V &= \frac{F_{\text{B}}}{kC_{\text{A0}}(1-X_{\text{A}})} = \frac{333.3\ \frac{\text{mol}}{\text{h}}}{\left(\frac{18}{1\ \text{h}}\right)\left(0.2\ \frac{\text{mol}}{\text{L}}\right)(1-X_{\text{A}})} \\ &= \left(\frac{92.592}{1-X_{\text{A}}}\right)L \end{aligned}$$

The annual equivalent cost of the reactor is

$$\begin{aligned} \$120V &= (\$120)\left(\frac{92.592}{1-X_{\text{A}}}\right) \\ &= \boxed{\$\left(\frac{11\,111}{1-X_{\text{A}}}\right)/\text{yr}} \end{aligned}$$

The answer is B.

2.3. The molar flow rate of the reactant required is

$$F_{\text{A0}} = \frac{F_{\text{B}}}{X_{\text{A}}} = \frac{333.33\ \frac{\text{mol A}}{\text{h}}}{X_{\text{A}}}$$

The annual cost of reactant is

$$\left(\frac{333.33}{X_A}\ \frac{\text{mol A}}{\text{h}}\right)\left(6000\ \frac{\text{h}}{\text{yr}}\right)\left(\frac{\$0.85}{\text{mol A}}\right)$$

$$= \boxed{\$\left(\frac{1.7\times 10^6}{X_A}\right)/\text{yr}}$$

The answer is A.

2.4. The annual operating cost is

$$\left(\frac{\$25}{1\ \text{h}}\right)\left(6000\ \frac{\text{h}}{\text{yr}}\right) = \$1.5\times 10^5/\text{yr}$$

The total annual cost is

$$\begin{aligned} T &= \text{annual cost of reactor} + \text{annual cost of reactant} \\ &\quad + \text{annual operating cost} \\ &= \$\left(\frac{11\,111}{1-X_A} + \frac{1.7\times 10^6}{X_A} + 1.5\times 10^5\right)/\text{yr}\end{aligned}$$

The cost of the reactant decreases with increasing conversion. However, higher conversion requires a larger reactor, which increases the cost of the reactor. Hence, these two competing cost factors need to be optimized by minimizing the total annual cost.

For total annual costs to be minimum, dT/dX_A must equal 0.

$$\frac{11\,111}{(1-X_A)^2} - \frac{1.7\times 10^6}{X_A^2} = 0$$

$$\left(\frac{1-X_A}{X_A}\right)^2 = \frac{11\,111}{1.7\times 10^6} = 0.0065$$

$$X_A = \boxed{0.9254}$$

The answer is D.

2.5. The optimum conversion is 92.54%. At this conversion the volume is

$$V = \frac{92.592}{1-X_A} = \frac{92.592}{1-0.9254} = \boxed{1241.2\ \text{L}}$$

The answer is C.

2.6. The molar flow rate of the reactant (feed) is

$$F_{A0} = \frac{333.33\ \frac{\text{mol}}{\text{h}}}{X_A} = \frac{333.33\ \frac{\text{mol}}{\text{h}}}{0.9254} = \boxed{360.20\ \text{mol A/h}}$$

The answer is B.

2.7. The total annual cost is

$$\begin{aligned} T &= \$\left(\frac{11\,111}{1-X_A} + \frac{1.7\times 10^6}{X_A} + 1.5\times 10^5\right)/\text{yr} \\ &= \$\left(\frac{11\,111}{1-0.9254} + \frac{1.7\times 10^6}{0.9254} + 1.5\times 10^5\right)/\text{yr} \\ &= \$2.136\times 10^6/\text{yr}\end{aligned}$$

The annual production of B is

$$\left(15\ \frac{\text{kg B}}{\text{h}}\right)\left(6000\ \frac{\text{h}}{\text{yr}}\right) = 90\,000\ \text{kg B/yr}$$

The cost per kilogram of B is

$$\frac{\frac{\$2.136\times 10^6}{\text{yr}}}{\frac{90\,000\ \text{kg B}}{\text{yr}}} = \boxed{\$23.75/\text{kg B}}$$

The answer is C.

SOLUTION 3

3.1. A schematic of the surge tank is shown.

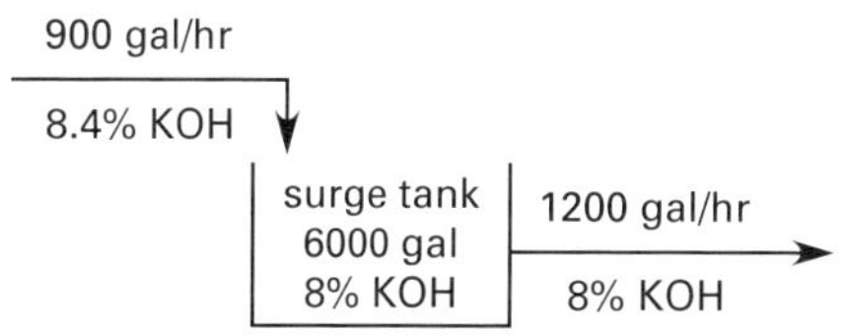

900 gal/hr
8.4% KOH
surge tank
6000 gal
8% KOH
1200 gal/hr
8% KOH

The volume of the liquid in the tank is

$$V = (6000\ \text{gal})\left(\frac{0.1337\ \text{ft}^3}{1\ \text{gal}}\right) = 802.2\ \text{ft}^3$$

For a cylindrical tank,

$$V = \left(\frac{\pi d^2}{4}\right)h$$

$$h = \frac{4V}{\pi d^2} = \frac{(4)\left(802.2\ \text{ft}^3\right)}{\pi\,(10\ \text{ft})^2} = \boxed{10.21\ \text{ft}}$$

The answer is D.

3.2. To determine the time required to empty the tank, a mass balance must be performed on the tank under transient conditions. The fundamental material balance equation for a nonreactive transient process is

$$\text{input} - \text{output} = \text{accumulation}$$

$$\dot{m}_{\text{in}} - \dot{m}_{\text{out}} = \frac{dm}{dt}$$

$$\dot{m}_{\text{in}} = \left(900\ \frac{\text{gal}}{\text{hr}}\right)\left(8.6\ \frac{\text{lbm}}{\text{gal}}\right) = 7740\ \text{lbm/hr}$$

$$\dot{m}_{\text{out}} = \left(1200\ \frac{\text{gal}}{\text{hr}}\right)\left(8.6\ \frac{\text{lbm}}{\text{gal}}\right) = 10{,}320\ \text{lbm/hr}$$

$$7740\ \frac{\text{lbm}}{\text{hr}} - 10{,}320\ \frac{\text{lbm}}{\text{hr}} = \frac{dm}{dt}$$

$$dm = -2580dt$$

The mass of the solution initially present in the tank (at $t = 0$) is

$$(6000\ \text{gal})\left(8.6\ \frac{\text{lbm}}{\text{gal}}\right) = 51{,}600\ \text{lbm}$$

Integrating,

$$\int_{51{,}600\ \text{lbm}}^{o} dm = -2580 \int_{o}^{t} dt$$

$$t = \frac{51{,}600\ \text{lbm}}{2580\ \frac{\text{lbm}}{\text{hr}}} = 20\ \text{hr}$$

The time required to empty the tank is 20 hr.

The answer is D.

3.3.

$$\dot{m}_{\text{in}} = \rho\dot{V}_{\text{in}}$$

$$\dot{m}_{\text{out}} = \rho\dot{V}_{\text{out}}$$

$$m = \rho V$$

$$\rho\dot{V}_{\text{in}} - \rho\dot{V}_{\text{out}} = \frac{d}{dt}(\rho V)$$

Since the fluid is incompressible, ρ is constant.

$$\dot{V}_{\text{in}} - \dot{V}_{\text{out}} = \frac{dV}{dt} = \left(\frac{\pi d^2}{4}\right)\frac{dh}{dt}$$

$$\frac{dh}{dt} = \frac{4\left(\dot{V}_{\text{in}} - \dot{V}_{\text{out}}\right)}{\pi d^2}$$

$$\dot{V}_{\text{in}} = \left(900\ \frac{\text{gal}}{\text{hr}}\right)\left(\frac{1\ \text{hr}}{60\ \text{min}}\right)\left(\frac{0.1337\ \text{ft}^3}{1\ \text{gal}}\right) = 2.006\ \text{ft}^3/\text{min}$$

$$\dot{V}_{\text{out}} = \left(1200\ \frac{\text{gal}}{\text{hr}}\right)\left(\frac{1\ \text{hr}}{60\ \text{min}}\right)\left(\frac{0.1337\ \text{ft}^3}{1\ \text{gal}}\right) = 2.674\ \text{ft}^3/\text{min}$$

$$\frac{dh}{dt} = \frac{(4)\left(2.006\ \frac{\text{ft}^3}{\text{min}} - 2.674\ \frac{\text{ft}^3}{\text{min}}\right)}{\pi\,(10\ \text{ft})^2} = -0.0085\ \text{ft/min}$$

$$dh = -0.0085dt$$

Integrating,

$$\int_{10.21\ \text{ft}}^{4\ \text{ft}} dh = -0.0085\int_0^t dt$$

$$t = \left(\frac{4\ \text{ft} - 10.21\ \text{ft}}{-0.0085\ \frac{\text{ft}}{\text{min}}}\right)\left(\frac{1\ \text{hr}}{60\ \text{min}}\right) = \boxed{12.17\ \text{hr}}$$

The answer is A.

3.4. The concentration of KOH in the inlet stream drops instantaneously to 6% because of the valve malfunction. For the alarm to sound, the KOH concentration in the outlet stream must drop to 7.8%. Perform a mass balance for KOH. Since the tank is well mixed, the concentration of KOH in the tank = concentration of KOH in the outlet solution = C. Let K = KOH.

Applying the material balance equation for KOH,

$$(\dot{m}_K)_{\text{in}} - (\dot{m}_K)_{\text{out}} = \frac{dm_K}{dt}$$

At the time of the valve malfunction,

$$(\dot{m}_K)_{\text{in}} = \dot{m}_{\text{in}}C_{\text{in}} = \left(7740\ \frac{\text{lbm}}{\text{hr}}\right)\left(0.06\ \frac{\text{lbm KOH}}{\text{lbm}}\right) = 464.4\ \text{lbm KOH/hr}$$

$$(\dot{m}_K)_{\text{out}} = \dot{m}_{\text{out}}C_{\text{out}} = \left(10{,}320\ \frac{\text{lbm}}{\text{hr}}\right)C = 10{,}320C\ \text{lbm KOH/hr}$$

$$\frac{dm_K}{dt} = \frac{d}{dt}(mC) = m\frac{dC}{dt} + C\frac{dm}{dt}$$

But,

$$\frac{dm}{dt} = -2580\ \text{lbm/hr}$$

Integrating,

$$\int_{51{,}600\ \text{lbm}}^{m} dm = -2580\int_0^t dt$$

$$m - 51{,}600\ \text{lbm} = -2580t$$

$$m = 51{,}600\ \text{lbm} - 2580t$$

(Note that m is in lbm and t is in hr.)

$$\frac{dm_K}{dt} = (51{,}600 - 2580t)\frac{dC}{dt} + C(-2580)$$

$$\begin{aligned}\frac{dm_K}{dt} &= (\dot{m}_K)_{\text{in}} - (\dot{m}_K)_{\text{out}} \\ &= 464.4\ \frac{\text{lbm KOH}}{\text{hr}} - 10{,}320C\ \frac{\text{lbm KOH}}{\text{hr}}\end{aligned}$$

$$\begin{aligned}464.4\ \frac{\text{lbm KOH}}{\text{hr}} - 10{,}320C\ \frac{\text{lbm KOH}}{\text{hr}} \\ = (51{,}600 - 2580t)\frac{dC}{dt} - 2580C\ \frac{\text{lbm KOH}}{\text{hr}}\end{aligned}$$

$$\begin{aligned}464.4\ \frac{\text{lbm KOH}}{\text{hr}} - 7740C\ \frac{\text{lbm KOH}}{\text{hr}} \\ = (51{,}600 - 2580t)\frac{dC}{dt}\end{aligned}$$

Separating the variables and integrating,

$$\begin{aligned}\int_{0.08}^{0.078} \frac{dC}{464.4 - 7740C} \\ = \int_0^t \frac{dt}{51{,}600 - 2580t}\end{aligned}$$

$$\begin{aligned}\left(-\frac{1}{7740}\right)\ln(464.4 - 7740C)\Big|_{0.08}^{0.078} \\ = \left(-\frac{1}{2580}\right)\ln(51{,}600 - 2580t)\Big|_0^t\end{aligned}$$

$$\begin{aligned}\ln\left(\frac{464.4 - 603.7}{464.4 - 619.2}\right) \\ = 3\ln\left(\frac{51{,}600 - 2580t}{51{,}600}\right)\end{aligned}$$

$$\ln(0.8999) = 3\ln(1 - 0.05t)$$

$$(1 - 0.05t)^3 = 0.8999$$

Solving for t,

$$t = \boxed{0.6909\ \text{hr}\quad(42\ \text{min})}$$

The answer is D.

3.5. From the solution to Prob. 3.4, the variation with time of the concentration of KOH in the outlet stream (as well as in the tank) is given by

$$\frac{dC}{464.4 - 7740C} = \frac{dt}{51{,}600 - 2580t}$$

Integrating between concentration limits of 0.08 and 0.07,

$$\int_{0.08}^{0.07} \frac{dc}{464.4 - 7740C} = \int_0^t \frac{dt}{51{,}600 - 2580t}$$

$$\ln\left(\frac{464.4 - 541.8}{464.4 - 619.2}\right) = 3\ln\left(\frac{51{,}600 - 2580t}{51{,}600}\right)$$

$$0.5 = (1 - 0.05t)^3$$

Solving for t,

$$t = \boxed{4.126\ \text{hr}}$$

The answer is A.

3.6. Start with the differential equation.

$$\frac{dC}{464.4 - 7740C} = \frac{dt}{51{,}600 - 2580t}$$

Integrating between time limits of 0 hr and 6 hr,

$$\int_{0.08}^{C} \frac{dC}{464.4 - 7740C} = \int_{0\ \text{hr}}^{6\ \text{hr}} \frac{dt}{51{,}600 - 2580t}$$

$$\ln\left(\frac{464.4 - 7740C}{464.4 - 619.2}\right) = 3\ \ln\left(\frac{36{,}120}{51{,}600}\right)$$

$$\left(\frac{464.4 - 7740C}{-154.8}\right) = 0.3430$$

$$C = \boxed{0.0669\quad(6.7\%)}$$

The answer is D.

SOLUTION 4

4.1. The design equation for a CSTR written in terms of component W is

$$\boxed{V = \frac{F_{\text{W}0}X_{\text{W}}}{-r_{\text{W}}}}$$

The answer is B.

4.2. $F_{\text{W}0}$ is the molar flow rate of component W entering the reactor.

$$\begin{aligned}F_{\text{W}0} = C_{\text{W}0}Q_0 &= \left(1.5\times10^{-4}\ \frac{\text{lbmol}}{\text{gal}}\right)\left(1500\ \frac{\text{gal}}{\text{hr}}\right) \\ &= \boxed{0.2250\ \text{lbmol/hr}}\end{aligned}$$

The answer is A.

4.3.

$$\begin{aligned} C_W &= C_{W0}(1 - X_W) \\ &= \left(1.5 \times 10^{-4} \ \frac{\text{lbmol}}{\text{gal}}\right)(1 - 0.97) \\ &= \boxed{4.5 \times 10^{-6} \ \text{lbmol/gal}} \end{aligned}$$

The answer is C.

4.4.

$$\begin{aligned} C_{O_2} &= 2.5 \times 10^{-6} \ \text{lbmol/gal} \\ k &= 43{,}500 \ \text{gal/lbmol-hr} \\ -r_W &= kC_W C_{O_2} \end{aligned}$$

$$\begin{aligned} V &= \frac{F_{W0}X_W}{-r_W} = \frac{F_{W0}X_W}{kC_W C_{O_2}} \\ &= \frac{\left(0.2250 \ \frac{\text{lbmol}}{\text{hr}}\right)(0.97)}{\left(43{,}500 \ \frac{\text{gal}}{\text{lbmol-hr}}\right)\left(4.5 \times 10^{-6} \ \frac{\text{lbmol}}{\text{gal}}\right) \times \left(2.5 \times 10^{-6} \ \frac{\text{lbmol}}{\text{gal}}\right)} \\ &= \boxed{445{,}997 \ \text{gal}} \end{aligned}$$

The answer is A.

4.5. In this case, $Q_0 = 3500$ gal/hr.

$$\begin{aligned} F_{W0} &= C_{W0}Q_0 = \left(1.5 \times 10^{-4} \ \frac{\text{lbmol}}{\text{gal}}\right)\left(3500 \ \frac{\text{gal}}{\text{hr}}\right) \\ &= 0.5250 \ \text{lbmol/hr} \\ V &= 445{,}997 \ \text{gal} \\ -r_W &= kC_W C_{O_2} = kC_{W0}(1 - X_W)C_{O_2} \end{aligned}$$

Substituting into the reactor design equation,

$$V = \frac{F_{W0}X_W}{kC_{W0}(1 - X_W)C_{O_2}}$$

$$445{,}997 \ \text{gal} = \frac{\left(0.5250 \ \frac{\text{lbmol}}{\text{hr}}\right)X_W}{\left(43{,}500 \ \frac{\text{gal}}{\text{lbmol-hr}}\right) \times \left(1.5 \times 10^{-4} \ \frac{\text{lbmol}}{\text{gal}}\right)(1 - X_W) \times \left(2.5 \times 10^{-6} \ \frac{\text{lbmol}}{\text{gal}}\right)}$$

Solving for X_W

$$X_W = \boxed{0.9327 \quad (93\% \text{ conversion})}$$

The answer is D.

4.6. The schematic for the two tanks operating in series is shown.

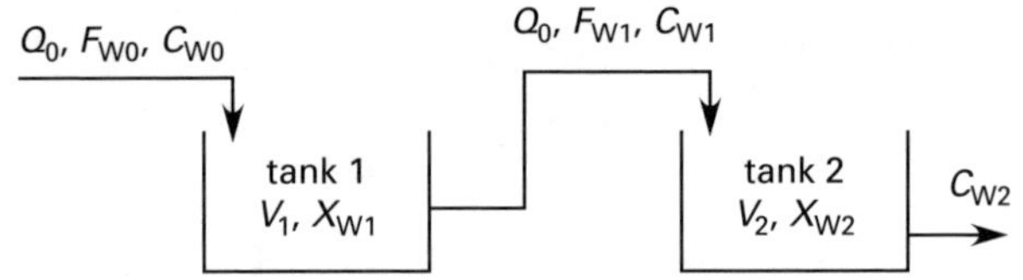

The overall conversion for the two tanks operating in series is 99%.

$$\begin{aligned} C_{W2} &= C_{W0}(1 - X_{\text{overall}}) \\ &= \left(1.5 \times 10^{-4} \ \frac{\text{lbmol}}{\text{gal}}\right)(1 - 0.99) \\ &= 1.5 \times 10^{-6} \ \text{lbmol/gal} \\ C_{W1} &= C_{W0}(1 - X_{W1}) \\ &= \left(1.5 \times 10^{-4} \ \frac{\text{lbmol}}{\text{gal}}\right)(1 - 0.9327) \\ &= 1.01 \times 10^{-5} \ \text{lbmol/gal} \\ C_{W2} &= C_{W1}(1 - X_{W2}) \end{aligned}$$

$$1.5 \times 10^{-6} \ \frac{\text{lbmol}}{\text{gal}} = \left(1.01 \times 10^{-5} \ \frac{\text{lbmol}}{\text{gal}}\right)(1 - X_{W2})$$

$$X_{W2} = \boxed{0.8515 \quad (85\% \text{ conversion})}$$

The answer is A.

4.7. The design equation for tank 2 is

$$V_2 = \frac{F_{W1}X_{W2}}{-r_{W2}}$$

$$\begin{aligned} F_{W1} &= C_{W1}Q_0 = \left(1.01 \times 10^{-5} \ \frac{\text{lbmol}}{\text{gal}}\right)\left(3500 \ \frac{\text{gal}}{\text{hr}}\right) \\ &= 0.0354 \ \text{lbmol/hr} \end{aligned}$$

$$-r_{W2} = kC_{W2}C_{O_2}$$

Since the temperature in tank 2 is the same as the temperature in tank 1, the value of k remains unchanged. The oxygen concentration is also identical in both the tanks. Substituting the values in the design equation gives

$$\begin{aligned} V_2 &= \frac{F_{W1}X_{W2}}{-r_{W2}} = \frac{F_{W1}X_{W2}}{kC_{W2}C_{O_2}} \\ &= \frac{\left(0.0354 \ \frac{\text{lbmol}}{\text{hr}}\right)(0.8515)}{\left(43{,}500 \ \frac{\text{gal}}{\text{lbmol-hr}}\right)\left(1.5 \times 10^{-6} \ \frac{\text{lbmol}}{\text{gal}}\right) \times \left(2.5 \times 10^{-6} \ \frac{\text{lbmol}}{\text{gal}}\right)} \\ &= \boxed{184{,}785 \ \text{gal}} \end{aligned}$$

The answer is B.

4.8. The molar flow rate of W leaving the second tank is F_{W2}.

$$F_{W2} = C_{W2}Q_0 = \left(1.5 \times 10^{-6}\ \frac{\text{lbmol}}{\text{gal}}\right)\left(3500\ \frac{\text{gal}}{\text{hr}}\right)$$

$$= \boxed{5.25 \times 10^{-3}\ \text{lbmol/hr}}$$

The answer is D.

4.9. The rate expression is

$$-r_W = kC_WC_{O_2}$$

The reaction is first order with respect to W but still depends on the oxygen concentration (which is maintained constant), making the reaction $\boxed{\text{pseudo-first order.}}$

The answer is D.

Turn to PPI for Your Chemical PE Exam Review Materials

The Most Trusted Source for Chemical PE Exam Preparation

Visit www.ppi2pass.com today!

The Most Comprehensive Reference Materials

Chemical Engineering Reference Manual for the PE Exam
Michael R. Lindeburg, PE

- ✔ The most widely used chemical PE exam reference
- ✔ 66 chapters provide in-depth review of exam topics
- ✔ More than 380 solved example problems
- ✔ Hundreds of key tables, charts, and graphs at your fingertips
- ✔ Full glossary for quick reference
- ✔ Quickly locate information through the complete index

Quick Reference for the Chemical Engineering PE Exam
Michael R. Lindeburg, PE

- ✔ Quickly and easily access the formulas needed most often during the exam
- ✔ Drawn from the *Chemical Engineering Reference Manual*
- ✔ Quickly retrieve formulas without the distraction of surrounding text
- ✔ Organized by topic and indexed for rapid retrieval

The Practice You Need to Succeed

Practice Problems for the Chemical Engineering PE Exam
Michael R. Lindeburg, PE

- ✔ The perfect companion to the *Chemical Engineering Reference Manual*
- ✔ 450 practice problems increase your problem-solving skills
- ✔ Multiple-choice format, just like the exam
- ✔ Coordinated with the *Chemical Engineering Reference Manual* for focused preparation
- ✔ Step-by-step solutions provide immediate feedback

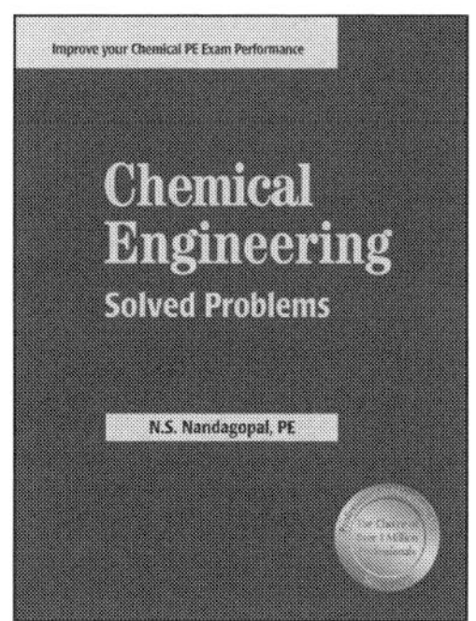

Chemical Engineering Solved Problems
N.S. Nandagopal, PE

- ✔ Collection of over 160 problems based on 26 scenarios
- ✔ Step-by-step solutions included for each problem
- ✔ Increase your speed and confidence
- ✔ Assess your strengths and weaknesses

Six-Minute Solutions for Chemical PE Exam Problems
Marta Vasquez, PhD, PE, and Robert R. Zinn

- ✔ Learn to solve problems in under 6 minutes
- ✔ 100 multiple-choice problems with solutions
- ✔ Improve your problem-solving speed and skills
- ✔ Perfect for the breadth and depth portions
- ✔ Discover how to avoid common mistakes

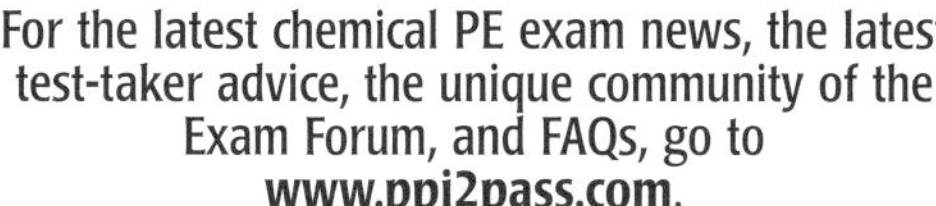

For the latest chemical PE exam news, the latest test-taker advice, the unique community of the Exam Forum, and FAQs, go to **www.ppi2pass.com**.